Digital Signal Processing

Digital Signal Processing:
DSP and Applications

Dag Stranneby

Newnes

OXFORD AUCKLAND BOSTON JOHANNESBURG MELBOURNE NEW DELHI

Newnes
An imprint of Butterworth-Heinemann
Linacre House, Jordan Hill, Oxford OX2 8DP
225 Wildwood Avenue, Woburn, MA 01801–2041
A division of Reed Educational and Professional Publishing Ltd

 A member of the Reed Elsevier plc group

First published 2001

British Library Cataloguing in Publication Data
A catalogue record for this book is available from the British Library

ISBN 0 7506 48112

Typeset by Florence Production Ltd, Stoodleigh, Devon
Printed and bound in Great Britain

Contents

Preface

This book is not a basic digital signal processing textbook. Even though the first chapter contains a brief summary of basic digital signal processing theory, it is assumed you have good knowledge of sampling, difference equations, z-transforms, FIR and IIR filters, FFT etc. The main idea of this book is to combine the topics of signal processing **theory** and implementing DSP algorithms in **practice**. The goal is also to give a broad **overview** of different digital signal processing applications, as a kind of appetizer for further studies.

Since it is my sincere belief that bringing **intuitive understanding** of concepts and systems is by far the most important part of teaching, the text is somewhat simplified at the expense of mathematical rigour. At the end of this book, references for deeper studies of the different topics covered can be found. Some details might be difficult for you to grasp at once, especially if you have very basic knowledge of DSP and its techniques. However, there is no cause for alarm. The most important first step of studying any subject is to grasp the **overall picture** and to understand the basics before delving into the details.

This book was originally written to be used in undergraduate courses at Örebro University in Sweden, but it has also been used for company on-site training. Besides the material covered in the following chapters, practical hands-on projects are an integral part of these courses. The projects consist of the design and implementation of signal processing algorithms on a digital signal processing system. Most of the programming work is done at assembly language level for gaining maximum understanding of both hardware and software problems.

Finally, I want to express my gratitude to Amy Loutfi for proof-reading the manuscript, checking the equations and figures and contributing many good ideas to this book.

Dag Stranneby

1 Introduction

Since the Second World War, if not earlier, technicians have speculated on the applicability of digital techniques to perform signal processing tasks. For example, at the end of the 1940s, Shannon, Bode and other researchers at the Bell Telephone Laboratories discussed the possibility of using digital circuit elements to implement filter functions. At this time, there was unfortunately no appropriate hardware available. Hence, cost, size and reliability strongly favoured conventional, analog implementations.

During the middle of the 1950s, Professor Linville at MIT discussed digital filtering at graduate seminars. By then, control theory, based partly on works by Hurewicz had become established as a discipline, and sampling and its spectral effects were well understood. A number of mathematical tools such as the z-transform, which had existed since Laplace's time, were now used in the electronics engineering community. Technology at that point, however, was only able to deal with low-frequency control problems or low-frequency seismic signal processing problems. While seismic scientists made notable use of digital filter concepts to solve problems, it was not until the middle of the 1960s that a more formal theory of digital signal processing began to emerge. During this period, the advent of the silicon integrated circuit technology made complete digital systems possible, but still quite expensive.

The first major contribution in the area of digital filter synthesis was made by Kaiser at Bell Laboratories. His work showed how to design useful filters using the bilinear transform. Further, in about 1965 the famous paper by Cooley and Turkey was published. In this paper, FFT (fast Fourier transform), an efficient and fast way of performing the DFT (discrete Fourier transform), was demonstrated.

At this time, hardware better suited for implementing digital filters was developed and affordable circuits started to be commercially available. Long FIR (finite impulse response) filters could now be implemented efficiently, thereby becoming a serious competitor to the IIR (infinite impulse response) filters, having better passband properties for a given number of delays. At the same time, new opportunities emerged. It was now possible to achieve time-varying, adaptive and non-linear filters that cannot be built using conventional analog techniques. One such filter is the Kalman filter named after R. E. Kalman. The Kalman filter is a model-based filter that filters the signal according to its statistical rather than its spectral properties.

In the area of adaptive filters, B. Widrow is an important name, especially when talking about the least mean square (LMS) algorithm. Widrow has also made significant contributions in the area of neural networks as early as in the 1960s and 1970s.

Today, there are many commercial products around, utilizing the advantages of digital signal processing, namely:

- an essentially perfect reproducibility
- a guaranteed accuracy (no individual tuning and pruning necessary)
- well suited for large volume production.

To conclude this section, we will give some everyday examples where digital signal processing is encountered in one way or another. Applications can be divided into two classes. The first class consists of applications that **could** be implemented using ordinary analog techniques, but where the use of digital signal processing increases the performance considerably. The second class of applications are these that **require** the use of digital signal processing and cannot be built using entirely 'analog' methods.

1.1.1 Measurements and analysis

Digital signal processing traditionally has been very useful in the areas of measurement and analysis in two different ways. One way is in preconditioning the measured signal by rejecting the disturbing noise and interference. The other way is in interpreting the properties of collected data by, for instance, correlation and spectral transforms. In the area of medical electronic equipment, more or less sophisticated digital filters can be found in electrocardiograph (ECG) and electroencephalogram (EEG) equipment to record the weak signals in the presence of heavy background noise and interference.

As pointed out earlier, digital signal processing has historically been used in systems dealing with seismic signals due to the limited bandwidth of these signals. Digital signal processing has also proven to be very well suited for air and space measuring applications, e.g. analysis of noise received from outer space by radio telescopes or analysis of satellite data. Using digital signal processing techniques for analysis of radar and sonar echoes are also of great importance in both civilian as well as military contexts.

Another application is in navigational systems. In most GPS (global positioning system) receivers today, advanced digital signal processing techniques are employed to enhance resolution and reliability.

1.1.2 Telecommunications

Digital signal processing is heavily used in many telecommunications systems today. For instance, they are used in telephone systems for DTMF (dual-tone multi-frequency) signalling, echo cancelling of telephone lines and equalizers used in high-speed telephone modems. Further, error-correcting codes are used to protect digital signals from bit errors during transmission (or storing) and different data compression algorithms are utilized to reduce the number of data bits needed to represent a given amount of information.

Digital signal processing is also used in many contexts in cellular telephone systems, for instance speech coding in GSM (groupe speciale mobile or global system for mobile communication) telephones, modulators and demodulators, voice scrambling and other cryptographic devices. An application dealing with high frequency is the directive antenna having an electronically controlled beam. By using directive antennas at the base stations in a cellular system, the base station can 'point' at the mobile at all times, thereby reducing the transmitter power needed. This in turn increases the capacity of a fixed bandwidth system in terms of the number of simultaneous users per square kilometre, i.e. it increases the service level and the revenue for the system operator.

1.1.3 Audio and television

In most audio equipment today, such as CD (compact disc) players, DAT (digital audio tape) and DCC (digital compact cassette) recorders, digital signal processing is mandatory. This is also true for most modern studio equipment as well as more or less advanced synthesizers used in today's music production. Digital signal processing has also made a lot of new noise suppression and companding systems (e.g. Dolby™) attractive.

Digital methods are not only used for producing and storing audio information, but also for distribution. This could be between studios and transmitters, or even directly to the end-user such as in the DAB (digital audio broadcasting) system. Digital transmission is also used for broadcasting of TV signals. The HDTV (high definition television) systems are utilizing lot of digital image processing techniques. **Digital image processing** can be regarded as a special branch of digital processing having many things in common with digital signal processing, but dealing mainly with two-dimensional image signals. Digital image processing can be used for many tasks, e.g. restoring distorted or blurred images, morphing, data compression by image coding, identification and analysis of pictures and photos.

1.1.4 Automotive

In the automotive business, digital signal processing is often used for control purposes. Some examples are ignition and injection control systems, 'intelligent' suspension systems, 'anti-skid' brakes, climate control systems, 'intelligent' cruise controllers and airbag controllers.

There are also systems for speech recognition and speech synthesis being tested in automobiles. Just tell the car 'Switch on the headlights' and it will, and maybe it will give the answer: 'The right rear parking light is not working'. Experiments have also been performed with background noise cancellation in cars using adaptive digital filters, and radar assisted, more or less 'smart' cruise controllers.

To summarize: digital signal processing is here to stay ...

1.2 Digital signal processing basics

This book is not intended to be a tutorial in basic digital signal processing. There are already numerous good references available on the subject (Oppenheimer and Schafer, 1975; Rabiner and Gold, 1975; Mitra and Kaiser, 1993; Marven and Ewers, 1993; Denbigh, 1998; Lynn and Fuerst, 1994). The following section is only a brief summary of underlying important theories and ideas with which the reader should already be familiar.

1.2.1 Continuous and discrete signals

In this book we will mainly study systems dealing with signals that vary over time (temporal signals). In 'reality' a signal can take on an infinite amount of values and time can be divided into infinitely small increments. A signal of this type is **continuous in amplitude** and **continuous in time**. In everyday language such a signal is called an 'analog' signal.

Now, if we only present the signal at given instants of time, but still allow the amplitude to take on any value, we have a signal type that is **continuous in amplitude**, but **discrete in time**.

The third signal type is a signal that is defined at all times, but only allowed to take on values from a given set. Such a signal is described as **discrete in amplitude** and **continuous in time**.

The fourth type of signal is one that is defined at given instants of time, and only allowed to take on values from a given set. This signal is said to be **discrete in amplitude** and **discrete in time**. This type of signal is called a 'digital' signal, and is the type of signal we will mainly deal with in this book.

1.2.2 Sampling and reconstruction

The process of going from a signal being **continuous in time** to a signal being **discrete in time** is called **sampling**. Sampling can be regarded as multiplying the time-continuous signal $g(t)$ with a train of unit pulses $p(t)$ (see Figure 1.1).

$$g^{\#}(t) = g(t)p(t) = \sum_{n=-\infty}^{+\infty} g(nT)\delta(t - nT) \qquad (1.1)$$

where $g^{\#}(t)$ is the sampled signal. Since the unit pulses are either one or zero, the multiplication can be regarded as a pure switching operation.

The time period T between the unit pulses in the pulse train is called the **sampling period**. In most cases this period is constant, resulting in 'equidistant sampling'. There is however no theoretical demand for the sampling period to be constant. In some systems, many different sampling periods are used ('multi-rate sampling') (Åström and Wittenmark, 1984). Yet, in other applications, the sampling period may be a stochastic variable. This results in 'random sampling' (Papoulis, 1985) which complicates the analysis considerably. In most systems today, it is common to use one or more constant sampling periods. The sampling period T is related to the **sampling rate** or **sampling frequency** f_s such that

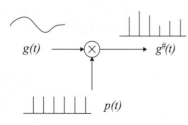

Figure 1.1 *Sampling viewed as a multiplication process*

$$f_s = \frac{\omega_s}{2\pi} = \frac{1}{T} \tag{1.2}$$

The process of sampling implies reduction of knowledge. For the time-continuous signal we know the value of the signal at every instant of time, but for the sampled version (the time-discrete signal) we only know the value at **specific points** in time. If we want to reconstruct the original time-continuous signal from the time-discrete sampled version, we therefore have to make more or less qualified interpolations of the values in between the sampling points. If our interpolated values differ from the true signal, we have introduced distortion in our reconstructed signal.

Now to get an idea of our chances of making a faithful reconstruction of the original signal, let us study the effect of the sampling process in the frequency domain. First, referring to equation (1.1), the pulse train $p(t)$ can be expanded in a Fourier series

$$p(t) = \sum_{n=-\infty}^{+\infty} c_n \, e^{jn(2\pi/T)t} \tag{1.3}$$

where the Fourier coefficients c_n are

$$c_n = \frac{1}{T} \int_{-T/2}^{+T/2} p(t) \, e^{-jn(2\pi/T)t} \, \mathrm{d}t = \frac{1}{T} \tag{1.4}$$

Hence, the sampling equation (1.1) can now be rewritten as

$$g^{\#}(t) = p(t)g(t) = \left(\frac{1}{T} \sum_{n=-\infty}^{+\infty} e^{jn(2\pi/T)t} \right) g(t) \tag{1.5}$$

The Laplace transform of the sampled signal $g^{\#}(t)$ is (using the multiplication-shift property)

$$G^{\#}(s) = \frac{1}{T} \sum_{n=-\infty}^{+\infty} G\left(s + jn \frac{2\pi}{T} \right) \tag{1.6}$$

To get an idea of what is happening in the frequency domain, we investigate equation (1.6) following the $s = j\omega$ axis

$$G^{\#}(j\omega) = \frac{1}{T} \sum_{n=-\infty}^{+\infty} G((j(\omega + n\omega_s)) \tag{1.7}$$

From this, we see that the effect of sampling creates an infinite number of copies of the spectrum of the original signal $g(t)$ Every copy is shifted by multiples of the sampling frequency ω_s. Figure 1.2 shows a part of the total spectrum.

The spectrum bandwidth of the original signal $g(t)$ is determined by the highest frequency component f_{max} of the signal. Now, two situations can occur. If $f_{max} < f_s/2$ then the copies of the spectrum will **not** overlap (Figure 1.2(a)). Given only one spectrum copy, we have a limited, but correct knowledge of the original signal $g(t)$ and can reconstruct it using an inverse Fourier transform. The sampling process constitutes an unambiguous mapping. If, on the other hand $f_{max} \geqslant f_s/2$, the spectra will overlap (Figure 1.2(b)) and the too-high frequency components will be aliased (or 'folded') into the lower part of the next spectrum. We can no longer reconstruct the original signal, since aliasing distortion has occurred. **Hence, it is imperative that the bandwidth of the original time-continuous signal being sampled is smaller than half the sampling frequency (also called the 'Nyquist' frequency).**

To avoid aliasing distortion in practical cases, the sampling device is always preceded by some kind of low-pass ('anti-aliasing') filter to reduce the bandwidth of the incoming signal. This signal is often quite complicated and contains a large number of frequency components. Since it is impossible to build perfect filters, there exists a risk of too-high frequency components leaking into the sampler, causing aliasing distortion. We also have to be aware that high-frequency interference may somehow enter the signal path after the low-pass filter, and we may experience aliasing distortion even though the filter is adequate.

In some literature the concept of 'relative frequency' (or 'fnosq') is used to make calculations simpler. The relative frequency is defined as

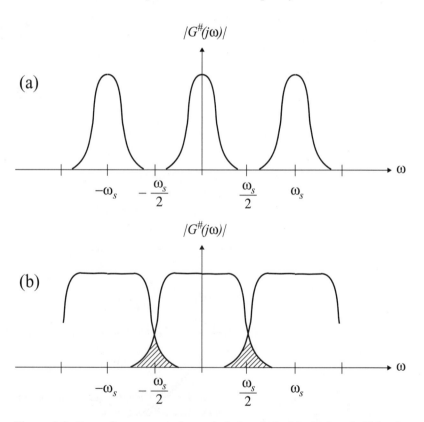

Figure 1.2 *Part of spectrum of sampled signal. In* (a) *the bandwidth of the signal is less than the Nyquist frequency $\omega_s/2$ and no aliasing takes place. In* (b) *the bandwidth is greater than $\omega_s/2$ and aliasing takes place, hence the original signal cannot be reconstructed*

$$q = \frac{f}{f_s} = \frac{\omega}{\omega_s} \tag{1.8}$$

Hence, to avoid aliasing distortion: $|q_{max}| < 0.5$

If the Nyquist criteria is met and hence no aliasing distortion is present, we can reconstruct the original bandwidth-limited time-continuous signal $g(t)$ in an unambiguous way. This is achieved using a low-pass **reconstruction filter** to extract only **one** copy of the spectrum from the sampled signal. It can be shown as a consequence of 'Shannon's sampling theorem' or the 'cardinal reconstruction formula' (Mitra and Kaiser, 1993) that the ideal low-pass filter to use for reconstruction has the impulse response of a sinc function

$$h(t) = \frac{\sin\left(\pi \frac{1}{T}\right)}{\pi \frac{t}{T}} = \operatorname{sinc}\left(\frac{1}{T}\right) \tag{1.9}$$

This is a **non-causal** filter (the present output signal depends on future input signals) having an infinite impulse response and an ideally sharp cutoff at the frequency $\pi/T = \omega_s/2$ radians per second, i.e. the Nyquist frequency. While of theoretical interest, this filter does not provide a practical way to reconstruct the sampled signal. A more realistic way to obtain a time-continuous signal from a set of samples is to hold the sample values of the signal constant between the sampling instants. This corresponds to a filter with an impulse response

$$h(t) = \begin{cases} 1/T & \text{if } 0 \leqslant t < T \\ 0 & \text{otherwise} \end{cases} \tag{1.10}$$

Such a reconstruction scheme is called a **zero-order hold** (ZOH) or **boxcar hold**, and it creates a 'staircase' approximation of the signal. The zero-order hold can be thought of as an approximation of the ideal reconstruction filter.

A more sophisticated approximation of the ideal reconstruction filter is the **linear point connector** or **first-order hold** (FOH), which connects sequential sample values with straight-line segments. The impulse response of this filter is

$$h(t) = \begin{cases} (T+t)/T^2 & \text{if } -\text{T} \leqslant t < 0 \\ (T-t)/T^2 & \text{if } \quad 0 \leqslant t < T \end{cases} \tag{1.11}$$

This filter provides a closer approximation to the ideal filter than the zero-order hold (Åström and Wittenmark, 1984).

1.2.3 Quantization

The sampling process described above is the process of converting a continuous time signal into a discrete time signal, while quantization is the conversion of a signal which is **continuous in amplitude** into a signal which is **discrete in amplitude**.

Quantization can be thought of as classifying the level of the continuous valued signal into certain bands. In most cases, these bands are equally spaced over a given range and undesired non-linear band spacing may cause harmonic distortion. Some applications (companding systems), use a non-linear band spacing which is often logarithmic.

Every band is assigned a code or numerical value. Once we have decided to which band the present signal level belongs, the corresponding code can be used to represent the signal level.

Most systems today use the binary code, i.e. the number of quantization intervals N is

$$N = 2^n \tag{1.12}$$

where n is the word length of the binary code. For example with $n = 8$ bits we get a resolution of $N = 256$ bands, $n = 12$ yields $N = 4096$ and $n = 16$ yields $N = 65536$. Obviously the more bands we have, i.e. the longer the word length, the better **resolution** we obtain. This in turn renders a more accurate representation of the signal.

Another way of looking at resolution of a quantization process is to define the **dynamic range** as the ratio between the strongest and the weakest signal level that can be represented. The dynamic range is often expressed in decibels. Since every new bit of word length being added increases the number of bands by a factor of 2, the corresponding increase in dynamic range is 6 dB. Hence an 8 bit system has a dynamic range of 48 dB, a 12 bit system 72 dB etc. (this of course only applies to linear band spacing).

Now, assume we have N bands in our quantizer, this implies that the normalized width of every band is $Q = 1/N$. Further, this means that a specific binary code will be presented for all continuous signal levels in the range of $\pm Q/2$ around the ideal level for the code. Hence, we have a random error in the discrete representation of the signal level of $\varepsilon = \pm Q/2$. This random error, being a stochastic variable, is independent of the signal and has a uniformly distributed probability density function ('rectangular' density function)

$$p(\varepsilon) = \begin{cases} 1/Q & \text{for } -Q/2 < \varepsilon < Q/2 \\ 0 & \text{else} \end{cases} \tag{1.13}$$

This stochastic error signal will be added to the 'true' discrete representation of our signal and appear as **quantization noise**. The RMS amplitude of the quantization noise can be expressed as (Pohlmann, 1989)

$$\varepsilon_{\text{rms}} = \sqrt{\int_{-Q/2}^{+Q/2} \varepsilon^2 p(\varepsilon)\, d\varepsilon} = \frac{Q}{\sqrt{12}} = \frac{2^{-n}}{\sqrt{12}} \tag{1.14}$$

The ratio between the magnitude of the quantization noise and the maximum signal magnitude allowed is denoted the **signal-to-error ratio**. A longer word length gives a smaller quantization noise and hence a better signal-to-error ratio.

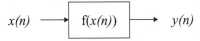

$x(n) \longrightarrow$ f(x(n)) $\longrightarrow y(n)$

Figure 1.3 *A discrete time linear digital signal processing operation, input x (n), output y (n) and transfer function f (x (n))*

1.2.4 Processing models for discrete time series

Assume that by proper sampling (at a constant sampling period *T*) and quantization we have now obtained a 'digital' signal $x(n)$. Our intention is to apply some kind of linear signal processing operation, for instance a filtering operation, to the signal $x(n)$ thereby obtaining a digital output signal $y(n)$ (see Figure 1.3).

The processing model is said to be **linear** if the **transfer function**, i.e. $y = f(x)$, the function defining the relationship between the output and input signals, satisfies the **principle of superposition**, such that

$$f(x_1 + x_2) = f(x_1) + f(x_2) \tag{1.15}$$

For clarity, let us illustrate this using a few simple examples. If *k* is a constant, the function (a perfect amplifier or attenuator)

$$f(x) = kx \tag{1.16}$$

is obviously linear since

$$f(x_1 + x_2) = k(x_1 + x_2) = kx_1 + kx_2 = f(x_1) + f(x_2) \tag{1.17}$$

Now, if we add another constant *m* (bias) what about the function

$$f(x) = kx + m \tag{1.18}$$

It looks linear, doesn't it? Well, let's see

$$f(x_1 + x_2) = k(x_1 + x_2) + m = kx_1 + kx_2 + m \tag{1.19a}$$

but

$$f(x_1) + f(x_2) = kx_1 + m + kx_2 + m = kx_1 + kx_2 + 2m \tag{1.19b}$$

Obviously, the latter function is **not** linear. Now we can formulate one requirement on a linear function: it has to pass through the origin, i.e. $f(0) = 0$.

If we consider a multiplier 'circuit' having two inputs y_1 and y_2, this can be expressed as a function having two variables

$$f(y_1, y_2) = y_1 y_2 \tag{1.20}$$

Let us now connect two composite signals $y_1 = x_1 + x_2$ and $y_2 = x_3 + x_4$ to each of the inputs. We now try to see whether the function satisfies the principle of superposition (equation 1.15) i.e. whether it is linear or not

$$f(x_1 + x_2, x_3 + x_4) = (x_1 + x_2)(x_3 + x_4) = x_1 x_3 + x_1 x_4 + x_2 x_3 + x_2 x_4 \tag{1.21a}$$

while

$$f(x_1, x_3) + f(x_2, x_4) = x_1 x_3 + x_2 x_4 \tag{1.21b}$$

Hence, a multiplication of two variables (composite signals) is a **non-linear** operation. For the special case we connect the same signal to both inputs, i.e. $y_1 = y_2$ we get

$$f(y_1, y_2) = y_2^2$$

which is obviously a non-linear function. If, on the other hand, we multiply a signal $y_1 = x_1 + x_2$ by a **constant** $y_2 = k$, we get

$$f(x_1 + x_2, k) = (x_1 + x_2)k$$

and

$$f(x_1, k) + f(x_2, k) = x_1 k + x_2 k$$

as in equations (1.16) and (1.17); then the operation is **linear**. A two-input function that is linear in each input if the other is held constant is called a **bilinear** function. Usually we call it a **product**.

The next observation we can make is that the first derivative $f'(x)$ of the function $f(x)$ is required to be a constant for the function to be linear. This also implies that higher order derivatives do not exist.

An arbitrary function $f(x)$, can be approximated using a Taylor or MacLaurin series

$$f(x) \approx f(0) + xf'(0) + \frac{x^2}{2!}f''(0) + \frac{x^3}{3!}f'''(0) \dots \tag{1.22}$$

By inspection of the first two terms we can find out whether the function is linear or not. To be linear we demand (see above)

$$f(0) = 0$$

and

$$f'(x) = k$$

It might be interesting to note that many common signal processing operations are indeed non-linear: rectifying, quantization, power estimation, modulation, demodulation, mixing signals (frequency translation), correlating etc. Filtering a signal using a filter with **fixed** coefficients is linear, while using an adaptive filter, having variable coefficients, may be regarded as a non-linear operation.

If the parameters of the transfer function are constant over time, the signal processing operation is said to be **time invariant**. Sometimes such an operation is referred to as an **LTI (linear time invariant) processor** (Lynn and Fuerst, 1994; Chen, 1999). Quite often such processors are also assumed to be **causal**, i.e. the present output signal only depends on present and past input signals.

The operation of a time-discrete processor can be expressed using a number of different mathematical models, for example, a difference equation model, a state–space model, a convolution model or a z-transform model.

The difference equation, describing the behaviour of a time-discrete system, can be regarded as the cousin of the differential equation describing a continuous-time ('analog') system. An example difference equation of the order of 2 is

$$y(n) - 0.8y(n-1) + 0.2y(n-2) = 0.1x(n-1) \tag{1.23}$$

To obtain a first output $y(0)$, the initial conditions $y(-1)$ and $y(-2)$ have to be known. The difference equation can of course be solved stepwise, by inserting the proper values of the input signal $x(n)$. In some cases this type of direct solution may be appropriate. It is however far more useful to obtain a closed form expression for the solution. Techniques for obtaining such closed forms are well described in the literature on difference equations (Mitra and Kaiser, 1993; Spiegel, 1971). The general form of the difference equation model is

$$\sum_{i=0}^{N} a_i y(n-i) = \sum_{j=0}^{M} b_j x(n-j) \tag{1.24}$$

where $x(n)$ is given and $y(n)$ is to be found. The order of the difference equation equals N, which is also the number of initial conditions needed to obtain an unambiguous solution.

   The state–space model can be seen as an alternative form of the difference equation, making use of matrix and vector representation. If, for example, we introduce the variables

$$\begin{cases} y_1(n) = y(n) \\ y_2(n) = y(n+1) \end{cases} \quad \text{and} \quad \begin{cases} x_1(n) = x(n) \\ x_2(n) = x(n+1) \end{cases} \tag{1.25}$$

our example difference equation (1.23) can be rewritten as

$$y_2(n-1) = 0.8y_2(n-2) - 0.2y_1(n-2) + 0.1x_2(n-2) \tag{1.26}$$

or in matrix equation form

$$\begin{bmatrix} y_1(n-1) \\ y_2(n-1) \end{bmatrix} = \begin{bmatrix} 0 & 1 \\ -0.2 & 0.8 \end{bmatrix} \begin{bmatrix} y_1(n-2) \\ y_2(n-2) \end{bmatrix} + \begin{bmatrix} 0 & 0 \\ 0 & 0.1 \end{bmatrix} \begin{bmatrix} x_1(n-2) \\ x_2(n-2) \end{bmatrix} \tag{1.27}$$

If we introduce the vectors

$$\mathbf{Y}(n) = \begin{bmatrix} y_1(n) \\ y_2(n) \end{bmatrix} \quad \text{and} \quad \mathbf{X}(n) = \begin{bmatrix} x_1(n) \\ x_2(n) \end{bmatrix}$$

and the matrices

$$\mathbf{A} = \begin{bmatrix} 0 & 1 \\ -0.2 & 0.8 \end{bmatrix} \quad \text{and} \quad \mathbf{B} = \begin{bmatrix} 0 & 0 \\ 0 & 0.1 \end{bmatrix}$$

The difference equation can now be rewritten in the compact matrix equation form

$$\mathbf{Y}(n+1) = \mathbf{A}\mathbf{Y}(n) + \mathbf{B}\mathbf{X}(n) \tag{1.28}$$

This is a very useful way of expressing the system. Having knowledge of the system state $\mathbf{Y}(n)$ and the input signal vector $\mathbf{X}(n)$ at time instant n, the new state of the system $\mathbf{Y}(n+1)$ at instant $n+1$ can be calculated. The system is completely specified by the transition matrix $\mathbf{A}$ and the input matrix $\mathbf{B}$. The state–space model is common when dealing with control system applications (Åström and Wittenmark, 1984; Chen, 1999).

The convolution model expresses the relationship between the input and output signals as a convolution sum

$$y(n) = \sum_{k=-\infty}^{+\infty} h(k)x(n-k) \tag{1.29}$$

If we choose the input signal $x(n)$ as the unit pulse $\delta(n)$, the output from the system will be

$$y(n) = \sum_{k=-\infty}^{+\infty} h(k)\delta(n-k) = h(n) \tag{1.30}$$

Hence, $h(n)$ is the **impulse response** of the system. If $h(n)$ of the system is known (or obtained by applying a unit pulse), the output $y(n)$ can be calculated for a given input signal $x(n)$.

Under some circumstances, the properties of the system in the **frequency domain** is of primary interest. These properties can be found by studying the system for a special class of input signals which are functionally equivalent to a sampled sinusoid of frequency ω

$$x(n) = e^{j\omega n} = \cos(\omega n) + j\sin(\omega n) \quad \text{for } -\infty < n < +\infty \tag{1.31}$$

Applying equation (1.31) to equation (1.29) we obtain

$$y(n) = \sum_{k=-\infty}^{+\infty} h(k)\,e^{j\omega(n-k)} = e^{j\omega n}\sum_{k=-\infty}^{+\infty} h(k)\,e^{-j\omega k} = x(n)H(e^{j\omega}) \tag{1.32}$$

Thus, for this special class of inputs, we see from equation (1.32) that the output is identical to the input to within a complex frequency dependent gain factor $H(e^{j\omega})$, which is defined from the impulse response of the system as

$$\frac{y(n)}{x(n)} = H(e^{j\omega}) = \sum_{k=-\infty}^{+\infty} h(k)\,e^{-j\omega k} \tag{1.33}$$

This gain factor is often referred to as the **frequency response** of the system. Taking the magnitude and argument of the frequency response, Bode-type plots showing gain and phase angle can be obtained. Note that the frequency response $H(e^{j\omega})$ is essentially the DFT (discrete fourier transform) of the impulse response $h(n)$. A well-known fact can be seen; that **a convolution in the time domain corresponds to a multiplication in the frequency domain** and vice versa.

Another way of representing the signal processing system is to use the *z*-**transform model**. If we assume that we are dealing with a causal system

$$h(n) = 0 \quad \text{and} \quad x(n) = 0 \quad \text{for } n < 0$$

or in other words, the input signal is zero for 'negative time' sampling instants, the summation interval of equation (1.29) can be reduced to

$$y(n) = \sum_{k=0}^{+\infty} h(k)x(n-k) \tag{1.34}$$

Taking the z-transform (Oppenheimer and Schafer, 1975; Rabiner and Gold, 1975; Denbigh, 1998), i.e. multiplying by z^{-n} and summing over $0 \leqslant n < \infty$ both sides of the equation (1.34) we obtain

$$\sum_{n=0}^{+\infty} y(n)z^{-n} = \sum_{n=0}^{+\infty}\sum_{k=0}^{+\infty} h(k)x(n-k)z^{-n} = \sum_{k=0}^{+\infty} h(k)z^{-k} \sum_{n=0}^{+\infty} x(n-k)z^{-(n-k)}$$

$$(1.35a)$$

$$Y(z) = H(z)X(z) \qquad\qquad (1.35b)$$

where the z-transform of the impulse response is

$$H(z) = \sum_{k=0}^{+\infty} h(k)z^{-k} \qquad\qquad (1.36)$$

Hence, the causal system can be fully characterized by the **transfer function** $H(z)$ which in the general case is an infinite series in the polynomial z^{-k}. For many series it is possible to find a closed-form summation expression.

Note for a system to be stable and hence useful in practice, the series must converge to a finite sum as $k \to \infty$.

When working in the z-plane, a multiplication by z^{-k} is equivalent to a delay of k steps in the time domain. This property makes, for instance, z-transformation of a difference equation, easy. Consider our example of the difference equation (1.23). If we assume that the system is causal and that the z-transforms $X(z)$ and $Y(z)$ of $x(n)$ and $y(n)$ respectively exist, it is straightforward to transform (1.23) as

$$Y(z) - 0.8Y(z)z^{-1} + 0.2Y(z)z^{-2} = 0.1X(z)z^{-1} \qquad (1.37)$$

Rearranging equation (1.37) we obtain

$$H(z) = \frac{Y(z)}{X(z)} = \frac{0.1z^{-1}}{1 - 0.8z^{-1} + 0.2z^{-2}} = \frac{0.1z}{z^2 - 0.8z + 0.2} \qquad (1.38)$$

The **zeros** are the roots of the numerator, while the **poles** are the roots of the denominator. Hence, in equation (1.38) above, the zero is

$$0.1z = 0$$

having the root $z_1 = 0$

and the poles are

$$z^2 - 0.8z + 0.2 = 0$$

having the complex roots $z_1 = 0.4 + j0.2$ and $z_2 = 0.4 - j0.2$

The study of the locations of the poles and zeros in the complex z-plane are sometimes referred to as the **root locus**. For the system described by equation (1.38) to be stable, it is required that the poles of the system lie within the unit circle in the complex z-plane, hence

$$|z_p| < 1 \quad \text{for all } p$$

For our example system above,

$$|z_1| = |z_2| = \sqrt{(0.4^2 + 0.2^2)} \approx 0.45$$

i.e. the system is stable.

In this section, we have demonstrated a number of ways to mathematically represent a signal processing operation (Figure 1.3). It is convenient to work with these models 'off line' in a powerful PC, workstation or minicomputer, using floating point arithmetic. It is often a challenge to migrate the system into a single-chip, fixed point, digital signal processor required to perform in real time. 'Smart' algorithms and 'tricks' are often needed to keep up processing speed and to avoid numerical truncation and overflow problems.

1.2.5 The non-recursive filter

This filter (Figure 1.4) is sometimes called a **tapped delay line filter**, a **transversal filter** or a **FIR filter** since it has a **finite impulse response**. Using the convolution model, the response can be expressed as

$$y(n) = \sum_{j=0}^{M} h(j)x(n-j) \tag{1.39}$$

where M is the length of the filter. If we apply a unit pulse to the input of the system, i.e. $x(n) = \delta(n)$, it is straightforward to realize that the impulse response $h(j)$ of the system is directly obtained from the weight (gain) of the taps of the filter

$$h(j) = b_j \tag{1.40}$$

If we prefer a z-plane representation of equation (1.39), one of the nice features of the FIR filter is apparent

$$H(z) = b_0 + b_1 z^{-1} + b_2 z^{-2} \ldots b_M z^{-M} \tag{1.41}$$

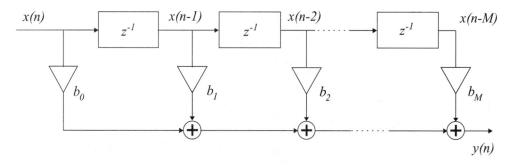

Figure 1.4 *Non-recursive (FIR) filter having length M with weights b_j*

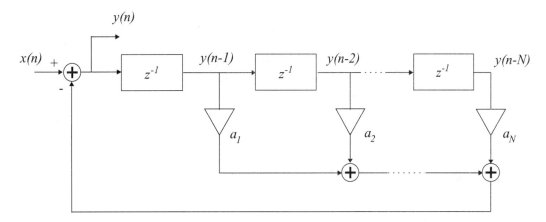

Figure 1.5 *Recursive (IIR) filter having length N with weights a_i*

Since the FIR filter does not have any poles, it is always guaranteed to be stable. Another advantage is that if the weights are chosen to be symmetrical (Lynn and Fuerst, 1994), the filter has a **linear phase response**, i.e. all frequency components experience the same time delay through the filter. There is no risk of distortion of compound signals due to phase shift problems. Further, knowing the amplitude of the input signal $x(n)$, it is easy to calculate the maximum amplitude of the signals in different parts of the system. Hence numerical overflow and truncation problems can easily be eliminated at design time.

The drawback with the FIR filter is that if sharp cutoff filters are needed so is a high-order FIR structure which results in long delay lines. FIR filters having hundreds of taps are however not uncommon today, thanks to low cost integrated circuit technology and high-speed digital signal processors.

1.2.6 The recursive filter

Recursive filters, sometimes called **IIR** filters use a feedback structure (Figure 1.5), and have an **infinite impulse response**. Borrowing some ideas from control theory, an IIR filter can be regarded as an FIR filter inserted in a feedback loop (Figure 1.6). Assume that the FIR filter has the transfer function $G(z)$; the transfer function of the total feedback structure, i.e. the IIR filter is then

Figure 1.6 *The IIR filter seen as an FIR filter in a feedback loop*

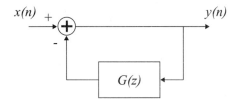

$$H(z) = \frac{1}{1 + G(z)} = \frac{1}{1 + a_1 z^{-1} + a_2 z^{-2} + \dots a_N z^{-N}} \tag{1.42}$$

The IIR filter only has poles, hence it is of great importance to choose the weights a_i in such a way that the poles stay inside the unit circle, to make sure the filter is stable. Since the impulse response is infinite, incoming samples will be 'remembered' by the filter. For this reason, it is not easy to calculate the amplitude of the signals inside the filter in advance, even if the amplitude of the input signal is known. Numerical overflow problems may occur in practise. The advantage of the IIR filter is that it is quite easy to build filters with sharp cutoff properties, using only a few delay elements.

Pure IIR filters are commonly used for ac coupling and smoothing (averaging) but it is more common to see combinations of FIR and IIR filter structures. Such a structure is the second-order combination shown in Figure 1.7, having the transfer function

$$H(z) = \frac{b_0 + b_1 z^{-1} + b_2 z^{-2}}{1 + a_1 z^{-1} + a_2 z^{-2}} \tag{1.43}$$

Of course, higher order structures than two can be built if required by the passband specifications. It is however more common to combine a number of structures of the order two in cascade or parallel to achieve advanced filters. In this way, it is easier to obtain numerical stability and to spot potential numerical problems.

Note! If a filter has zeros **cancelling** the poles (i.e. having the same position in the complex plane), the total transfer function may have only zeros. In such a case the filter will have a finite impulse response (FIR), despite

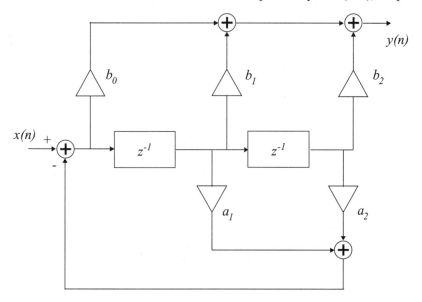

Figure 1.7 *Combined second order FIR and IIR filter structure. Note: only two delay elements are needed*

the fact that the structure may be of the recursive type and an infinite impulse response (IIR) is expected. Hence, a non-recursive (FIR) filter **always** has a finite impulse response, while a recursive filter commonly has an infinite impulse response, but may in some cases have a finite impulse response.

If, for instance, a non-recursive filter is transformed into a recursive structure, the recursive filter will have the same impulse response as the original non-recursive filter, i.e. a FIR. Consider for example a straightforward non-recursive averaging filter having equal weights

$$y(n) = \frac{1}{N} \sum_{i=0}^{N-1} x(n - i) \tag{1.44}$$

The filter has the transfer function

$$H_N(z) = \frac{Y(z)}{X(z)} = \frac{1}{N} \sum_{i=0}^{N-1} z^{-i} \tag{1.45}$$

Obviously, this filter has a finite impulse response. To transform this FIR filter into a recursive form, we try to find out how much of the previous output signal $y(n-1)$ can be reused to produce the present output signal $y(n)$

$$y(n) - y(n-1) = \frac{1}{N} \sum_{i=0}^{N-1} x(n-i) - \frac{1}{N} \sum_{i=0}^{N-1} x(n-i-1)$$

$$= \frac{1}{N} x(n) - \frac{1}{N} x(n-N) \tag{1.46}$$

The corresponding recursive filter structure can hence be written

$$y(n) = y(n-1) + \frac{1}{N} x(n) - \frac{1}{N} x(n-N)$$

$$= y(n-1) + \frac{1}{N} (x(n) - x(n-N)) \tag{1.47}$$

having the transfer function

$$H_R(z) = \frac{Y(z)}{X(z)} = \frac{1}{N} \cdot \frac{1 - z^{-N}}{1 - z^{-1}} \tag{1.48}$$

This is indeed a recursive filter having the same impulse response as the first filter and hence a FIR. Another interesting observation concerns the implementation of the filter. In the first case, one multiplication and $N-1$ additions are needed, while in the second case only one multiplication, one subtraction and one addition is required to perform the same task. Since digital filters are commonly implemented as computer programs, less computational burden implies faster execution and shorter processing time. In this example, the recursive algorithm seems advantageous, but this is not generally true.

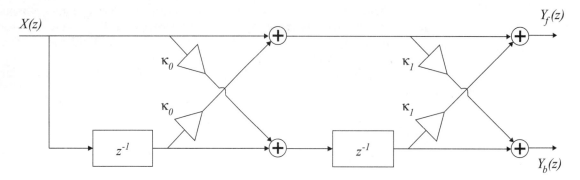

Figure 1.8 *An example all-zero lattice filter consisting of two lattice elements*

1.2.7 The lattice filter

Beside the standard filter structures discussed above, there are many specialized ones for specific purposes. One example of such a structure is the **lattice filter** (Orfandis, 1985; Widrow and Stearns, 1985) which is commonly used in adaptive processing and particularly in linear prediction. An example lattice filter consisting of two lattice elements is shown in Figure 1.8. There are many variations of lattice structures. This example is an all-zero version, i.e. a structure having a transfer function containing only zeros (no poles), so it may be regarded as a lattice version of a non-recursive filter (FIR). The filter has one input and two outputs. The upper output signal is often denoted the **forward prediction error**, while the lower output signal is called the **backward prediction error**. The reason for this will be explained below.

The transfer function, with respect to the forward prediction error, has the general form

$$H_f(z) = \frac{Y_f(z)}{X(z)} = \sum_{i=0}^{L} b_i z^{-i} \tag{1.49}$$

where $b_0 = 1$ and L is the number of lattice elements in the filter. In our example $L = 2$, hence

$$Y_f(z) = X(z) + X(z) b_1 z^{-1} + X(z) b_2 z^{-2} \tag{1.50}$$

The corresponding difference equation is

$$y_f(n) = x(n) + b_1 x(n-1) + b_2 x(n-2) = x(n) - \hat{x}(n) \tag{1.51}$$

Now, if we assume that we can find filter coefficients b_i in such a way that $y_f(n)$ is small or preferably equal to zero for all n, we can interpret $\hat{x}(n)$ as the **prediction** of the input signal $x(n)$ based on $x(n-1)$ and $x(n-2)$ in this example. Hence, we are able to predict the input signal one step **forward** in time. Further, under these assumptions it is easy to see that $y_f(n) = x(n) - \hat{x}(n)$ is the forward prediction error.

In a similar way, it can be shown that the transfer function with respect to the backward prediction error (the lower output signal) has the general form

$$H_b(z) = \frac{Y_b(z)}{X(z)} = \sum_{i=0}^{L} b_{L-i} z^{-i} \tag{1.52}$$

where $b_0 = 1$ and L is the number of lattice elements in the filter. In our example $L = 2$, hence

$$Y_b(z) = X(z)b_2 + X(z)b_1 z^{-1} + X(z)z^{-2} \tag{1.53}$$

The corresponding difference equation is

$$y_b(n) = x(n-2) + x(n)b_2 + b_1 x(n-1) = x(n-2) - \hat{x}(n) \tag{1.54}$$

In this case $\hat{x}(n)$ is the **prediction** of the input signal $x(n-2)$ based on $x(n-1)$ and $x(n)$. Hence, we are able to predict the input signal one step **backward** in time and $y_b(n) = x(n-2) - \hat{x}(n)$ is the backward prediction error.

The procedure of determining the lattice filter coefficients (Orfandis, 1985; Widrow and Stearns, 1985) $\varkappa_i$ is not entirely easy and will not be covered in this book. There are of course other types of filters which are able to predict signal sample values more than only one step ahead. There are also more advanced types, e.g. non-linear predictors. An important application using predictors is data compression. The idea is to use the dependency between present samples and future samples, thereby reducing the amount of information.

2 The analog–digital interface

2.1.1 General

In most systems, whether electronic, financial or social, most problems arise in the interface between different sub-parts. This is of course also true for digital signal processing systems. Most signals in real life are continuous in amplitude and time, i.e. 'analog', but our digital system is working with amplitude- and time-discrete signals, so-called 'digital' signals. Hence, the input signals entering our system need to be converted from analog to digital form before the actual signal processing may take place.

For the same reason, the output signals from our digital signal processing device need to be reconverted back from digital to analog form, to be used, for instance, in hydraulic valves or loudspeakers or other analog actuators. These conversion processes, between the analog and digital world add some problems to our system. These matters will be addressed in this chapter, together with a brief presentation of some common techniques used to perform the actual conversion processes.

2.1.2 Encoding and modulation

Assuming we have now converted our analog signals to numbers in the digital world, there are many ways to **encode** the digital information into the shape of electrical signals. This process is called **modulation** (sometimes 'line modulation'). The most common method is probably **pulse code modulation (PCM)**. There are two common ways of transmitting PCM, and these are **parallel** and **serial** mode. In an example of the parallel case, the information is encoded as voltage levels on a number of wires, called a parallel **bus**. We are using binary signals, which means that only two voltage levels are used, +5 V corresponding to a binary '1' (or 'true') and 0 V meaning a binary '0' (or 'false'). Hence, every wire carrying 0 V or +5 V contributes a binary digit ('bit'). A parallel bus consisting of 8 wires will hence carry 8 bits, a byte consisting of bits D0, D1 to D7 (Figure 2.1). Parallel buses are able to transfer high information data rates, since an entire data word, i.e. a sampled value, is being transferred at a time. This transmission can take place between for instance an analog-to-digital converter and a digital signal processor. One drawback with parallel buses is that they require a number of wires, i.e. board space, on a printed circuit board. Another

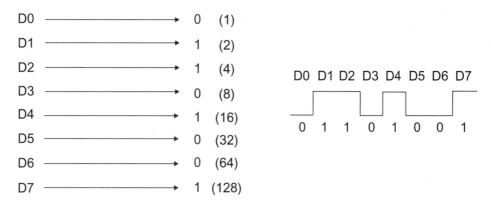

Figure 2.1 *Example, a byte (96H) encoded (weights in parenthesis) using PCM in parallel mode (parallel bus, 8 bits, 8 wires) and in serial mode as an 8 bit pulse train (over one wire)*

problem is that we may experience skew problems, i.e. different time delays on different wires, meaning that all bits will not arrive at the same time in the receiver end of the bus and data words will be messed up. Since this is especially true for long, high-speed parallel buses, this kind of bus is only suited for comparatively short transmission distances. Protecting long parallel buses from picking up wireless interference or to radiate interference may also be a formidable problem.

The alternative way of dealing with PCM signals is to use the **serial** transfer mode. In this case, the bits are not transferred on different wires in parallel, but in sequence on a single wire (see Figure 2.1). First, bit D0 is transmitted, then D1 etc. This of course means that the transmission of, for instance, a byte, requires a longer time than in the parallel case. On the other hand, only **one** wire is needed. Board space and skew problems will be eliminated and interference problems can be easier to solve.

There are many possible modulation schemes such as pulse amplitude modulation (PAM), pulse position modulation (PPM), pulse number modulation (PNM), pulse width modulation (PWM) and pulse density modulation (PDM). All these modulation types are used in serial transfer mode (see Figure 2.2).

Pulse amplitude modulation (PAM)
The actual amplitude of the pulse represents the number being transmitted. Hence, PAM is continuous in amplitude but discrete in time. The output of a sampling circuit with a zero-order hold is one example of a PAM signal.

Pulse position modulation (PPM)
A pulse of fixed width and amplitude is used to transmit the information. The actual number is represented by the position in time where the pulse appears in a given time slot.

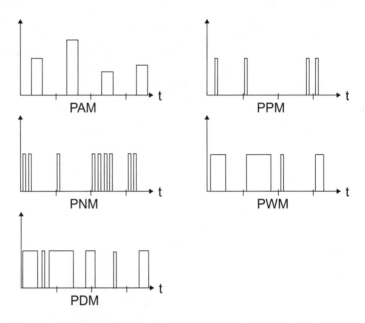

Figure 2.2 *Different modulation schemes for serial mode data communication, Pulse Amplitude Modulation, Pulse Position Modulation, Pulse Number Modulation, Pulse Width Modulation and Pulse Density Modulation*

Pulse number modulation (PNM)
Related to PPM in the sense that we are using pulses with fixed amplitude and width. In this modulation scheme however, many pulses are transmitted in every time slot, and the number of pulses present in the slot represents the number being transmitted.

Pulse width modulation (PWM)
Quite common modulation scheme, especially in power control and power amplifier contexts. In this case, the width (duration) T_1 of a pulse in a given time slot T represents the number being transmitted. If the pulse has the amplitude A_1, the transmitted number is represented by

$$A_1 \frac{T_1}{T} \tag{2.1}$$

In most applications, the amplitude A_1 of the pulse is fixed and uninteresting. Only the time ratio is used in the transmission process. If however, the amplitude of the pulse is also used to represent a second signal, we are using a combination of PAM and PWM. In some applications, this is a simple way of achieving a multiplication of two signals.

Pulse density modulation (PDM)
May be viewed as a type of degenerated PWM, in the sense that not only the width of the pulses changes, but also the periodicity (frequency). The

number being transmitted is represented by the density or 'average' of the pulses.

A class of variations of PDM called **stochastic representation of variables** was tried in the 1960s and 1970s. The idea was to make a 'stochastic computer', replacing the analog computer consisting of operational amplifiers and integrators, working directly with the analog signals. The stochastic representation has two nice features. Firstly, the resolution can be traded for time, which implies resolution can be improved by transmitting more pulses (longer time needed). Secondly, the calculations can be performed easily using only standard combinatorial circuits. The idea of stochastic representation experienced a renaissance in the 1980s in some forms of neural network applications.

So far we have talked about 'transmission' of digital information using different types of modulation. This discussion is of course also relevant for **storing** digital information. When it comes to optical (CD) or magnetic media, there are a number of special modulation methods (Pohlmann, 1989; Miller, 1998) used, which will not be treated here.

As will be seen later in this chapter, some signal converting and processing chips and sub-systems may use different modulation methods to communicate. This may be due to standardization or due to the way the actual circuit works. One example is the so-called **CODEC** (coder-decoder). For instance, this is a chip used in telephone systems, containing both an analog-to-digital converter and a digital-to-analog converter and other necessary functions to implement a full two-way analog/digital interface for voice signals. Many such chips use a serial PCM modulation interface. Switching devices and digital signal processors commonly have built-in interfaces to handle this type of signal.

2.1.3 Number representation and companding systems

When the analog signal is quantized, it is commonly represented by binary numbers in the following processing steps. There are many possible representations of quantized amplitude values. One way is to use **fixed-point** formats like **2's complement**, **offset binary** or **sign and magnitude** (Pires, 1989). Another way is to use some kind of **floating-point** format. The difference between the fixed point formats can be seen in Table 2.1.

The most common fixed-point representation is 2's complement. In the digital signal processing community, we often interpret the numbers as **fractions**, rather than integers. This will be discussed in subsequent chapters. Other codes (Pires, 1989) are Gray code and BCD (binary coded decimal).

There are a number of floating-point formats around. They all rely on the principle of representing a number in three parts: a sign bit, an exponent and a mantissa. One such common format is the **IEEE Standard 754.1985** single precision 32 bit format, where the floating point number is represented by one sign bit, an 8 bit exponent and a 23 bit mantissa. Using this method, numbers between $\pm 3.37 \cdot 10^{38}$ and $\pm 8.4 \cdot 10^{-37}$ can be represented using only 32 bits. Note however that the use of floating-point representation only expands the **dynamic range** at the expense of the resolution and system

Table 2.1 *Some fixed point binary number formats*

Integer	2's Complement	Offset binary	Sign and magnitude
7	0111	1111	0111
6	0110	1110	0110
5	0101	1101	0101
4	0100	1100	0100
3	0011	1011	0011
2	0010	1010	0010
1	0001	1001	0001
0	0000	1000	0000
−1	1111	0111	1000
−2	1110	0110	1001
−3	1101	0101	1010
−4	1100	0100	1011
−5	1011	0011	1100
−6	1010	0010	1101
−7	1001	0001	1110
−8	1000	0000	1111

complexity. For instance, a 32 bit fixed-point system may have better resolution than a 32 bit floating-point system, since in the floating-point case, the resolution is determined by the word length of the mantissa being only 23 bits. Another problem encountered when using the floating-point systems is the **signal-to-noise-ratio (SNR)**. Since the size of the quantization steps will change as the exponent changes, so will the quantization noise. Hence, there will be discontinuous changes in SNR at specific signal levels. In an audio system, audible distortion (Pohlmann, 1989) may result from the modulation and quantization noise created by barely audible low-frequency signals causing numerous exponent switches.

From the above, we realize that fixed point (linear) systems yields **uniform** quantization of the signal. Meanwhile floating-point systems, due to the range changing, provide a **non-uniform** quantization. Non-uniform quantization is often used in systems where a compromise between word length, dynamic range and distortion at low signal levels has to be found. By using larger quantization steps for larger signal levels and smaller steps for weak signals, a good dynamic range can be obtained without causing serious distortion at low signal levels or requiring unreasonable word lengths (number of quantization steps). A digital telephone system may serve as an example where small signal levels are the most probable ones, thus causing the need for good resolution at low levels to keep distortion low. On the other hand, sometimes stronger signals will be present, and distortion due to saturation is not desirable. Due to the large number of connections in a digital telephone switch, the word length must be kept low, commonly not more than 8 bits.

Another way of accomplishing non-uniform quantization is **companding** (Pohlmann, 1989), a method which is not being 'perfect', but easy to

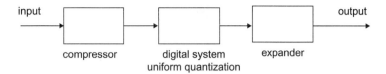

Figure 2.3 *A companding system consisting of a **compressor**, a common digital system using uniform quantization intervals and an **expander**. NOTE! The signal processing algorithms used must take compression and expansion into account*

implement. The underlying idea is to use a system utilizing common uniform quantization and reconstruction. At the input of the system a **compressor** is connected and at the output of the system an **expander** (hence, **comp**ressor/exp**ander**: '**compander**') is added (see Figure 2.3).

The compressor is mainly a non-linear amplifier, often logarithmic, having a lower gain for stronger signals than for weaker ones. In this way, the dynamic range of the input signal is compressed. The expander is another non-linear amplifier having a function being the inverse of the compressor. Hence, the expander 'restores' the dynamic range of the signal at the output of the system. The total system will now act as a system using non-uniform quantization. Note: the signal processing algorithms in the system must take the non-linearity of the compressor and expander into account.

For speech applications, typically telephone systems, there are two common non-linearities (Pohlmann, 1989) used: the **μ-law** and the **A-law**. The μ-law is usually used in the USA and the logarithmic compression characteristic has the form

$$y = \frac{\log(1 + \mu x)}{\log(1 + \mu)} \quad \text{for } x \geq 0 \tag{2.2}$$

where y is the magnitude of the output, x the magnitude of the input and μ a positive parameter defined to yield the desired compression characteristic. A parameter value of 0 corresponds to linear amplification, i.e. no compression and uniform quantization. A value of $\mu=255$ is often used to encode speech signals (see Figure 2.4). In such a system, an 8 bit implementation can achieve a good SNR and a dynamic range equivalent to that of a 12 bit system using uniform quantization. The inverse function is used for expansion.

The A-law is primarily used in Europe. Its compression characteristic has the form

$$y = \begin{cases} \dfrac{Ax}{1 + \log(A)} & \text{for } 0 \leq x < \dfrac{1}{A} \\[3ex] \dfrac{1 + \log(Ax)}{1 + \log(A)} & \text{for } \dfrac{1}{A} \leq x < 1 \end{cases}$$

$$\tag{2.3}$$

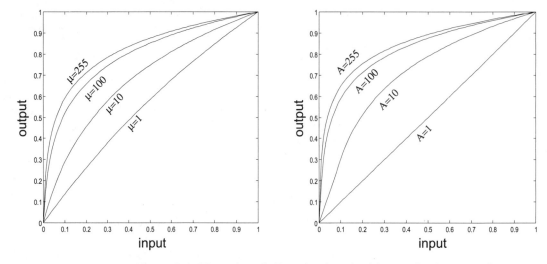

Figure 2.4 *The μ-law (left) and A-law (right) transfer functions for different parameter values*

where y is the magnitude of the output, x the magnitude of the input and A is a positive parameter defined to yield the desired compression characteristic.

Figure 2.4 shows μ-law and A-law transfer functions for some values of μ and A. Companding techniques are also used in miscellaneous noise reduction systems.

2.2 Digital to analog conversion

From now on we will only discuss systems using uniform quantization intervals. The task of the digital to analog converter (**DAC**) is to convert a numerical, commonly binary, so-called 'digital' value, into an 'analog' output signal. The DAC is subject to many requirements such as offset, gain, linearity, monotonicity and settling time.

Offset is the analog output when the digital input calls for a zero output. This should of course ideally be zero. The offset error affects all output signals with the same additive amount and in most cases it can be sufficiently compensated for by external circuits or by **trimming** the DAC.

Gain or scale factor, is the slope of the transfer curve from digital numbers to analog levels. Hence, the gain error is the error in the slope of the transfer curve. This error affects all output signals by the same percentage amount, and can normally be (almost) eliminated by trimming the DAC or by means of external circuitry.

Linearity can be sub-divided into integral linearity (relative accuracy) and differential linearity. **Integral linearity** error is the deviation of the transfer curve from a straight line. This error is not possible to adjust or compensate for easily.

Differential linearity measures the difference between any two adjacent output levels. If the output level for one step differs from the previous step by exactly the value corresponding to one LSB (least significant bit) of the digital value, the differential non-linearity is zero. Differential linearity errors cannot be eliminated easily.

Monotonicity implies that the analog output must increase as the digital input increases and decrease as the input decreases for all values over the specified signal range. Non-monotonicity is a result of excess differential non-linearity ($\geqslant$ 1 LSB). This implies that a DAC which has a differential non-linearity specification of a maximum of ± 0.5 LSB is more tightly specified than one for which only monotonicity is guaranteed. Monotonicity is essential in many control applications to maintain precision and to avoid instabilities in feedback loops.

Absolute accuracy error is the difference between the measured analog output from a DAC compared to the expected output for a given digital input. The absolute accuracy is the compound effect of the offset error, gain error and linearity errors described above.

Settling time of a DAC is the time required for the output to approach a final value within the limits of an allowed error band for a step change in the digital input. Measuring the settling time may be difficult in practice, since some DACs produce glitches when switching from one level to another. These glitches, being considerably larger than the fraction of the 1 LSB step of interest, may saturate, for instance, an oscilloscope input amplifier, thereby causing significant measuring errors. DAC settling time is a parameter of importance mainly in high sampling rate applications.

One important thing to remember is that the parameters above may be affected by supply voltage and temperature. In DAC data sheets, the parameters are only specified for certain temperatures and supply voltages, e.g. normal room temperature ($+25^\circ$ C) and nominal supply voltage. Considerable deviations from the specified figures may occur in a practical system.

2.2.1 Multiplying D/A converters

This is the most common form of DAC. The output is the product of an input current or reference voltage and an input digital code. The digital information is assumed to be in PCM parallel format. There are also DACs with a built-in shift register circuit, converting serial PCM to parallel. Hence, there are multiplying DACs for both parallel and serial transfer mode PCM available. Multiplying DACs have the advantage of being fast. In Figure 2.5 a generic current source multiplying DAC is shown. The bits in the input digital code are used to turn on a selection of current sources, which are then summed to obtain the output current. The output current can easily be converted into an output voltage using an operational amplifier.

Another way of achieving the different current sources in Figure 2.5 would

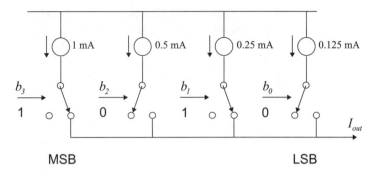

Figure 2.5 *A generic multiplying DAC using current sources controlled by the bits in the digital input code*

be to use a constant input reference voltage U_{ref} and a set of resistors. In this way, the currents for the different branches would simply be obtained by

$$I_i = \frac{U_{ref}}{R_i} b_i = \frac{U_{ref}}{2^{N-1-i} R_{N-1}} b_i \qquad (2.4)$$

where R_i is the resistance of the resistor in the i-th branch being controlled by the i-th bit: b_i being 1 or 0. The word length of the digital code is N and $i = 0$ is LSB. The total output current can then be expressed as the sum of the currents from the branches

$$I_{out} = \sum_{i=0}^{N-1} I_i = \frac{U_{ref}}{R_{N-1}} \sum_{i=0}^{N-1} \frac{b_i}{2^{N-1-i}} \qquad (2.5)$$

Building such a DAC in practice would however cause some problems. This is especially true as the word length increases. Assume for instance that we are to design a 14 bit DAC. If we choose the smallest resistor, i.e. the resistor R_{13} corresponding to $i = N - 1 = 13$ to be 100 ohms, the resistor R_0 will then need to be $R_0 = 2^{13} \cdot 100 = 819200$ ohms. Now, comparing the currents flowing in branch 0 and 13 respectively we find: $I_0 = U_{ref}/819200$ A and $I_{13} = U_{ref}/100$ A. To obtain good differential linearity, the error of current I_{13} (corresponding to the MSB (most significant bit)) must be smaller than the smallest current I_0 (corresponding to the LSB). Hence the resistance of R_{13} is required to be correct within ±122 ppm (parts per million). Another problem is that different materials and processes may be required when making resistors having high or low resistance. This will result in resistors in the DAC having different ageing and temperature stability properties, thus making it harder to maintain specifications under all working conditions.

It is possible to achieve the required precision by laser trimming, but there is a smarter and less expensive way to build a good DAC using an **R–2R ladder** structure. In an R–2R ladder, there are considerably more than N resistors, but they all have the same resistance R and are hence easier to manufacture and integrate on a single silicon chip. The R–2R ladder structure uses current division. A simple 3 bit R–2R ladder DAC is shown in Figure 2.6. All resistors R1 to R11 have the same resistance R. As can be

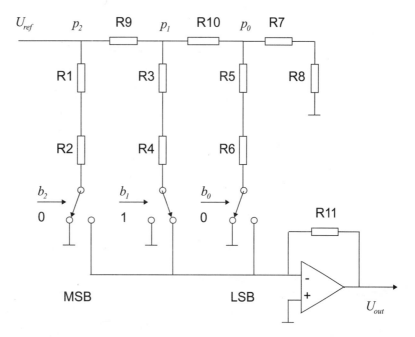

Figure 2.6 *An example 3 bit, R-2R ladder multiplying DAC, with voltage output, all resistors have the same resistance, R. Input digital code: 010*

seen from the figure, it is quite common that two resistors are connected in series, e.g. R1–R2, R3–R4, R5–R6 and R7–R8. Each of these pairs can of course be replaced by a single resistor having the resistance $2R$. That is why this structure is called R–2R ladder, it can be built simply by using two resistance values: R and $2R$.

The switches b_2, b_1, b_0 are controlled by bits 2, 1 and 0 in the digital code. If a bit is set to one, the corresponding switch is in its right position, i.e. switched to the negative input of the operational amplifier. If the bit is set to zero, the switch is in its left position, connecting the circuit to ground. The negative input of the operational amplifier is a current summation point, and due to the feedback amplifier having its positive input connected to ground, the negative input will be held at practically zero potential, i.e. ground ('virtual ground'). Hence, the resistor ladder circuit will be loaded in the same way, the position of the switches does not matter.

Now, let us examine the resistor ladder structure a bit more closely. Starting in junction point p_0 we conclude that the current flowing through resistor R10 will be divided into the two branches R5–R6 and R7–R8 respectively. As seen above, both branches have the same resistance namely $2R$. This means that the current will be divided equally in the two branches, i.e. half of the current passing through R10 will pass through the switch b_0. Further, if we calculate the total resistance of the circuit R5–R6 in parallel with R7–R8, we will find it to be R. If we now move to point p_1 we will have a similar situation as in p_0. The total resistance of R5, R6, R7 and R8 is R. This combination is connected in series with R10 having resistance

R ohms, hence the total resistance in the circuit passing through R10 to ground will be $2R$. Since the total resistance through R3–R4 is also $2R$, the current flowing through resistor R9 will be divided equally in the two branches as in the earlier case. One half of the current will pass through switch b_1 and one half will continue down the ladder via R10. For every step in the ladder, the same process is repeated. The current created by the reference voltage U_{ref} will be divided by two for every step in the ladder, and working with binary numbers, this is exactly what we want.

Assuming the operational amplifier to be an ideal one (zero offset voltage, zero bias current and infinite gain) the output voltage from the DAC can then be written

$$U_{out} = -\frac{U_{ref}}{2}\left(b_2 + \frac{b_1}{2} + \frac{b_0}{4}\right) = -\frac{U_{ref}}{2}\sum_{i=0}^{N-1}\frac{b_i}{2^{N-1-i}} \tag{2.6}$$

Another way of building a DAC is by utilizing **charge redistribution**. This technique is quite common in CMOS (complementary metal oxide semiconductor) single-chip computers and the DAC can be implemented on a silicon chip in the same way as, for instance, switched capacitor filters. This type of DAC makes use of the fact that if a fixed voltage U_{ref} is applied to a (variable) capacitor C, the charge in the capacitor will be

$$Q = CU_{ref} \tag{2.7}$$

After charging the capacitor C, it is disconnected from the voltage source and connected to another (discharged) capacitor C_1. The electric charge Q is then redistributed between the two capacitors and the voltage over the capacitors will be

$$U = \frac{Q}{C + C_1} = U_{ref}\frac{C}{C + C_1} \tag{2.8}$$

An example of a generic charge redistribution DAC is shown in Figure 2.7. The DAC works in two phases. In phase 1, the reset phase, all switches that are connected to ground close, discharging all capacitors. In phase 2, switch S is opened and the remaining switches are controlled by the incoming digital word. If a bit b_i is one, the corresponding capacitor $C_i = C/2^{N-1-i}$ is connected to U_{ref}, else it is connected to ground.

It is quite straightforward to see how the DAC works. In phase 2, when the capacitors are charged and the total capacitance C_{tot} as seen from the voltage source U_{ref} is

$$C_{tot} = \frac{C_{one}\left(C_{zero} + \dfrac{C}{2^{N-1}}\right)}{C_{one} + \left(C_{zero} + \dfrac{C}{2^{N-1}}\right)} \tag{2.9}$$

where C_{one} is the total capacitance of the capacitors corresponding to the bits that are 1 in the digital word (these capacitors will appear to be connected in parallel) and C_{zero} is the total capacitance of the capacitors corresponding

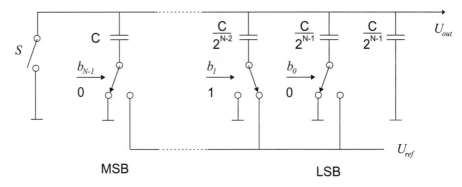

Figure 2.7 *A generic charge redistribution DAC*

to the bits that are 0. Hence the total charge of the circuit will be

$$Q_{tot} = C_{tot} U_{ref} \tag{2.10}$$

Now, the output voltage U_{out} is the voltage over the capacitors connected to ground

$$U_{out} = \frac{Q_{tot}}{\left(C_{zero} + \dfrac{C}{2^{N-1}}\right)} = \frac{U_{ref} C_{tot}}{\left(C_{zero} + \dfrac{C}{2^{N-1}}\right)} \tag{2.11}$$

Inserting equation (2.9) into equation (2.11) and expressing the capacitances as sums, we finally obtain the output voltage of the DAC as a function of the bits b_i in the digital code

$$U_{out} = \frac{U_{ref} C_{one} \left(C_{zero} + \dfrac{C}{2^{N-1}}\right)}{\left(C_{zero} + \dfrac{C}{2^{N-1}}\right)\left(C_{one} + \left(C_{zero} + \dfrac{C}{2^{N-1}}\right)\right)}$$

$$= \frac{U_{ref} C_{one}}{C_{one} + C_{zero} + \dfrac{C}{2^{N-1}}} \tag{2.12}$$

$$= \frac{U_{ref} C \displaystyle\sum_{i=0}^{N-1} \dfrac{b_i}{2^{N-1-i}}}{C \displaystyle\sum_{i=0}^{N-1} \dfrac{1}{2^{N-1-i}} + \dfrac{C}{2^{N-1}}} = U_{ref} \sum_{i=0}^{N} \frac{b_i}{2^{N-1-i}}$$

There are many alternative charge based circuits around. It is for instance possible to design a type of C–2C ladder circuit.

2.2.2 Integrating D/A converters

This class of DACs are also called **counting** DACs. These DACs are often slow compared to the previous type of converters. On the other hand, they may offer high resolution using quite simple circuit elements. No high precision resistors etc. are needed.

The basic building blocks are: an 'analog accumulator' usually called an integrator, a voltage (or current) reference source, an analog selector, a digital counter and a digital comparator. Figure 2.8 shows an example of an integrating DAC. The incoming N bits of PCM data (parallel transfer mode) are fed to one input of the digital comparator. The other input of the comparator is connected to the binary counter having N bits, counting pulses from a clock oscillator running at frequency f_c

$$f_c \geqslant 2^N f_s \tag{2.13}$$

where f_s is the sampling frequency of the system and N is the word length. Now, assume that the counter starts counting from 0. The output of the comparator will be zero and there will be a momentary logic one output from the comparator when the counter digital value equals the digital input PCM code. This pulse will set the bistable flip-flop circuit. The flip-flop will then be reset when the counter wraps around from $2^N - 1$ to 0 and a carry signal is generated.

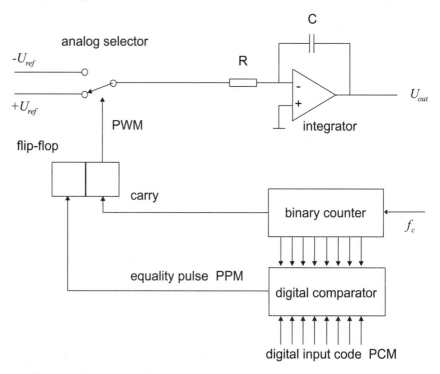

Figure 2.8 *An example integrating (counting) DAC (in a real world implementation, additional control and synchronization circuits are needed)*

The output from the flip-flop controls the analog selector in such a way that when the flip-flop is reset, $-U_{ref}$ is connected to the input of the integrator, and when the flip-flop is set, $+U_{ref}$ is selected. Note, that the output of the comparator is basically a PPM version of the PCM input data, and the output of the flip-flop is a PWM version of the same quantity. Hence, if we happen to have the digital code available in PWM or PPM format instead of parallel PCM, the circuit can be simplified accordingly.

The integrator simply averages the PWM signal presented to the input, thus producing the output voltage U_{out}. The precision of this DAC depends on the stability of the reference voltages, the performance of the integrator and the timing precision of the digital parts, including the analog selector.

There are many variants of this basic circuit. In some types, the incoming PCM data is divided into a 'high' and 'low' half, controlling two separate voltage or current reference selectors. The reference voltage controlled by the 'high' data bits is higher than the one controlled by the 'low' data bits. This type of converter is often denoted a **dual slope converter**.

2.2.3 Bitstream D/A converters

This type of DAC relies on the **oversampling** principle, i.e. using a considerably higher sampling rate than required by the Nyquist criteria. Using this method, sampling rate can be traded for accuracy of the analog hardware and the requirements of the analog reconstruction filter on the output can be relaxed. Oversampling reduces the problem of accurate N bit data conversion to a rapid succession of, for instance, 1 bit D/A conversions. Since the latter operation involves only an analog switch and a reference voltage source, it can be performed with high accuracy and linearity.

The concept of oversampling is to increase a fairly low sampling frequency to a higher one by a factor called the OSR (oversampling ratio). Increasing the sampling rate implies that more samples are needed than are available in the original data stream. Hence, 'new' sample points in between the original ones have to be created. This is done by means of an **interpolator**, also called an **oversampling filter** (Pohlmann, 1989). The simplest form of interpolator creates new samples by making a linear interpolation between two 'real' samples. In many systems, more elaborate interpolation functions are often used, implemented as a cascade of digital filters. As an example, an oversampling filter in a CD player may have 16 bit input samples at 44.1 kHz sampling frequency and an output of 28 bit samples at 176.4 kHz, i.e. an OSR of 4.

The interpolator is followed by the **truncator** or M bit **quantizer** (Figure 2.9). The task of the truncator is to reduce the number of N bits in the incoming data stream to M bits in the outgoing data stream ($N > M$). The truncation process is simply performed by taking the M most significant bits of the incoming N bits. This process however creates a strong quantization noise in the passband of interest. This is counteracted by means of the **noise shaping feedback loop**, consisting of a delay element and an adder. The $M-N$ least significant bits are fed back through the delay element and subtracted from the incoming data stream. Hence, the signal transfer func-

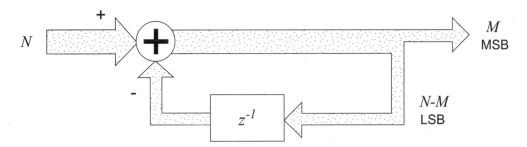

Figure 2.9 *A truncator with noise shaping feedback, input N bits, truncated to output M bits*

tion of the truncator equals 1 for the signal, but constitutes a high-pass filter having the transfer function

$$H(z) = \frac{1}{1 + z^{-1}} \tag{2.14}$$

for the quantization noise. The noise level in the interesting passband is attenuated, while at higher frequencies, the noise increases. The higher frequency noise components will however be attenuated by the analog reconstruction filter following the DAC. To conclude our example above, the CD player may then have a truncator with $N = 28$ bits and $M = 16$ bits. The sampling rate of both the input and output data streams are 176.4 kHz. In this example we still have to deal with 16 bits data in the DAC, but we have simplified the analog reconstruction filter on the output.

Oversampling could be used to derive yet one more advantage. Oversampling permits the noise shaper to transfer information in the 17th and 18th bits of the signal into a 16 bit output with duty cycle modulation of the 16th bit. In this way a 16 bit DAC can be used to convert an 18 bit digital signal. The 18 bit capability is retained because the information in the two 'surplus' bits is transferred in the four times oversampled signal (OSR = 4). Or, in other words, the averaged (filtered) value of the quarter-length 16 bit samples is as accurate as that of 18 bit samples at the original sampling rate.

Now, if the truncator is made in such a way that the output data stream is 1 bit, i.e. $M = 1$, this bitstream is a PDM version of the original N bit wide data stream. Hence, a one bit DAC can be used to convert the digital signal to its analog counterpart (see Figure 2.10).

To improve the resolution, a **dither** signal is added to the LSB. This is a pseudo random sequence, a kind of noise signal. Assume that the digital oversampled value lies somewhere in between two quantization levels. To obtain an increased accuracy, smaller quantization steps, i.e. an increased word length (more bits) is needed. Another way of achieving improved resolution is to add random noise to the signal. If the signal lies half way between two quantization levels, the signal plus noise will be equally probable to take on the high and low quantized value respectively. The actual level of the signal will hence be represented in a stochastic way. Due to the averaging process in the analog reconstruction low-pass filter, a better estimate

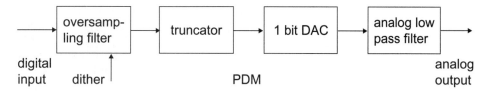

Figure 2.10 *A PDM bitstream 1 bit DAC, with (dither) pseudo noise added*

of the actual signal level will be obtained than without dither. If the signal lies closer to one quantization level than the other, the former quantized value will be more probable and hence, the average will approach this quantized level accordingly.

Dither signals are also used in control systems, including actuators like electrohydraulic valves. In this case, the dither signal may be a single sinus tone, superimposed on the actuator control signal. The goal is to prohibit the mechanics of the valve to 'hang up' or stick.

2.2.4 Sample-and-hold and reconstruction filters

The output from a DAC can be regarded as a PAM representation of the digital signal at the sampling rate. An ideal sample represents the value of the corresponding analog signal in a single **point** in time. Hence, in an ideal case, the output of a DAC is a train of impulses, each having an infinitesimal width, thus eliminating the **aperture error**. The aperture error is caused by the fact that a sample in a practical case does occupy a certain interval of time. The narrower the pulse width of the sample, the less the error. Of course, ideal DACs cannot be built in practice.

Another problem with real world DACs is that during the transition from one sample value to another, glitches, ringing and other types of interference may occur. To counteract this, a **sample-and-hold** device (S&H or S/H) is used. The most common type is the **zero-order hold** (ZOH) presented in Chapter 1. This device keeps the output constant until the DAC has settled on the next sample value. Hence, the output of the S/H is a staircase waveform approximation of the sampled analog signal. In many cases, the S/H is built into the DAC itself. Now, the S/H having a sample pulse impulse response, has a corresponding transfer function of the form

$$H(f) = \frac{\sin{(\pi f T)}}{\pi f T} = \text{sinc}\,(fT) \tag{2.15}$$

where T is the sampling rate, i.e. the holding time of the S/H. The function (2.15) represents a low-pass filter with quite mediocre passband properties. In Chapter 1 we concluded that the function required ideally to reconstruct the analog signal would be an ideal low-pass filter, having a completely flat passband and an extremely sharp cutoff at the Nyquist frequency. Obviously, the transfer function of the S/H is far from ideal. In many cases an analog **reconstruction filter** or **smoothing filter** (or anti-image filter) is needed in the signal path after the S/H, to achieve a good enough reconstruction of the analog signal. Since the filter must be implemented using analog components, it tends to be bulky and expensive and it is preferably kept simple, of the order of 3 or lower. A good way of relaxing the requirements of the filter is to use oversampling as described above. There are also additional requirements on the reconstruction filter depending on the application. In a high quality audio system, there may be requirements regarding linear phase shift and transient response, while in a feedback control system time delay parameters may be crucial.

In most cases a reconstruction filter is necessary. Even if a poorly filtered output signal has an acceptable quality, the presence of imaged high frequency signal components may cause problems further down the signal path. For example, assume that an audio system has the sampling frequency 44.1 kHz and the sampled analog sinusoidal signal has a frequency of $f_1 = 12$ kHz, i.e. well below the Nyquist frequency. Further, the system has a poor reconstruction filter at the output and the first high frequency image signal component $f_2 = 44.1 - f_1 = 44.1 - 12 = 32.1$ kHz leaks through. Now, the 32.1 kHz signal is not audible and does not primarily cause any problems. If however, there is a non-linearity in the signal path, for instance an analog amplifier approaching saturation, the signal components will be mixed. New signal components at lower frequencies may be created, thus causing audible distortion. For instance, a third-order non-linearity, quite common in bipolar semi-conductor devices, will create the following 'new' signal components due to intermodulation (mixing):

$$f_3 = 8.1 \text{ kHz} \qquad f_4 = 36.0 \text{ kHz} \qquad f_5 = 52.2 \text{ kHz}$$

$$f_6 = 56.1 \text{ kHz} \qquad f_7 = 76.2 \text{ kHz} \qquad f_8 = 96.3 \text{ kHz}$$

Note that the first frequency, 8.1 kHz, is certainly audible and will cause distortion. The high-frequency components can of course also interfere with e.g. bias oscillators in analog tape recorders and miscellaneous radio frequency equipment, causing additional problems.

Finally, again we find an advantage using oversampling techniques. If the sampling frequency is considerably higher than twice the Nyquist frequency, the distance to the first mirrored signal spectrum will be comparatively large (see Chapter 1). The analog filter now needs to cut off at a higher frequency, hence the gain can drop off more slowly with frequency and a simpler filter will be satisfactory.

2.3 Analog to digital conversion

The task of the analog to digital converter (**ADC**) is the 'inverse' of the DAC, i.e. to convert an 'analog' input signal into a numerical, commonly binary so called 'digital' value. The specifications for an ADC are similar to those for a DAC, i.e. offset, gain, linearity, missing codes, conversion time and so on.

Offset error is the difference between the analog input level which causes a first bit transition to occur and the level corresponding to 1/2 LSB. This should of course ideally be zero, i.e. the first bit transition should take place at a level representing exactly 1/2 LSB. The offset error affects all output codes with the same additive amount and can in most cases be sufficiently compensated for by adding an analog DC level to the input signal and/or by adding a fixed constant to the digital output.

Gain, or scale factor, is the slope of the transfer curve from analog levels to digital numbers. Hence, the gain error is the error in the slope of the transfer curve. It affects all output codes by the same percentage amount, and can normally be counteracted by amplification or attenuation of the analog input signal. Compensation can also be performed, by multiplying the digital number with a fixed-gain calibration constant.

As pointed out above, offset and gain errors can, to a large extent, be compensated for by preconditioning the analog input signal or by processing the digital output code. The first method requires extra analog hardware, but has the advantage of utilizing the ADC at its best. The digital compensation is often easier to implement in software, but cannot fully eliminate the occurrence of unused quantization levels and/or overloading of the ADC.

Linearity can be sub-divided into integral linearity (relative accuracy) and differential linearity.

Integral linearity error is the deviation of code midpoints of the transfer curve from a straight line. This error is not possible to adjust or compensate for easily.

Differential linearity measures the difference between input levels corresponding to any two adjacent digital codes. If the input level for one step differs from the previous step by exactly the value corresponding to one LSB, the differential non-linearity is zero. Differential linearity errors cannot be eliminated easily.

Monotonicity implies that increasing the analog input level never results in a decrease of the digital output code. Non-monotonicity may cause stability problems in feedback control systems.

Missing codes in an ADC mean that some digital codes can never be generated. This problem indicates that differential non-linearity is larger than 1 LSB. Missing codes are generally caused by a non-monotonic behaviour of the internal DAC. As will be shown in later sections, some types of ADCs use a built-in DAC.

Absolute accuracy error is the difference between the actual analog input to an ADC compared to the expected input level for a given digital output. The absolute accuracy is the compound effect of the offset error, gain error and linearity errors described above.

Conversion time of an ADC is the time required by the ADC to perform a complete conversion process. The conversion is commonly started by a 'strobe', or synchronization signal, controlling the sampling rate.

As for DACs, it is important to remember that the parameters above may be affected by supply voltage and temperature. Data sheets only specify the parameters for certain temperatures and supply voltages. Significant deviations from the specified figures may hence occur in a practical system.

2.3.1 Anti-aliasing filters and sample-and-hold devices

As pointed out in Chapter 1, the process of sampling the analog time-continuous signal requires an efficient **anti-aliasing filter** to obtain an unambiguous digital, time-discrete representation of the signal. Ideally, a low-pass filter having a flat passband and extremely sharp cutoff at the Nyquist frequency is required. Of course, building such a filter in practice is impossible and approximations and compromises thus have to be made. The problem is quite similar to building the 'perfect' reconstruction filter.

The first signal processing block of a digital signal processing system is likely to be an analog low-pass anti-aliasing filter. Depending on the application, different requirements on the filter may be stated. In audio systems, linear phase response may for instance be an important parameter, while in a digital DC-voltmeter instrument, a low offset voltage may be imperative. Designing proper anti-aliasing filters is, in the general case, not a trivial task, especially if practical limitations such as circuit board space and cost also have to be taken into account.

Anti-aliasing filters are commonly implemented as active filters using feedback operational amplifiers or as switched capacitor (SC) filters. One way to relax the requirements of the analog anti-aliasing filter is to use **oversampling** techniques (see also reconstruction filters for DAC above). In this case, the signal is sampled with a considerably higher rate than required to fulfill the Nyquist criteria. Hence, the distance to the first mirrored signal spectrum on the frequency axis will be much longer than if sampling were performed at only twice the Nyquist frequency. Figure 2.11 shows an example of an oversampled system. This could for instance be an audio system, where the highest audio frequency of interest would be about 20 kHz. The sampling rate for such a system could be $f_s = 48$ kHz. Without oversampling, the analog anti-aliasing filter should hence have a gain of about 1 at 20 kHz, and an attenuation of say 30 dB at 24 kHz. This corresponds to a slope of roughly 150 dB/octave, quite an impressive filter and very hard to implement using analog components.

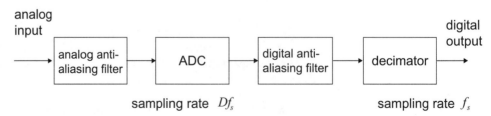

Figure 2.11 *D times oversampling ADC relaxing demands on the analog anti-aliasing filter*

Now, assume that we sample the analog input signal using $Df_s = 16 \times 48$ = 768 kHz, i.e. using an OSR of $D = 16$. To avoid aliasing distortion in this case, the analog anti-aliasing filter has quite relaxed requirements: gain of 1 at 20 kHz (as before) and attenuation of 30 dB at 384 kHz, which corresponds to a slope of 1.6 dB/octave. Such a filter is very easy to realize using a simple passive RC network.

Following the ADC, there is another anti-aliasing filter having the same tough requirements as stated previously. On the other hand, this filter is a digital one, being implemented using digital hardware or as software in a digital signal processor (DSP). Designing a digital filter having the required passband characteristics is not very hard. Preferably, a FIR filter structure may be used, having a linear phase response.

At last, following the digital anti-aliasing filter is a **decimator** or **downsampler**. To perform the downsampling (Pohlmann, 1989), the device only passes every D-th sample from input to output and ignores the others, hence

$$y(n) = x(Dn) \tag{2.16}$$

In our example the decimator only takes every 16th sample and passes it on, i.e. the sampling rate is now brought back to $f_s = 48$ kHz.

Another detail that needs to be taken care of when performing analog-to-digital conversion is the actual sampling of the analog signal. An ideal sampling implies measuring the analog signal level during an infinitely short period of time (the 'aperture'). Further, the ADC requires a certain time to perform the conversion and during this process the analog level must not change or conversion errors may occur. This problem is solved by feeding the analog signal to a **sample-and-hold** device (S/H) before it reaches the ADC. The S/H will take a quick snapshot sample and hold it constant during the conversion process. Many ADCs have a built-in S/H device.

2.3.2 Flash A/D converters

Flash-type (or parallel) ADCs are the fastest, because of the short conversion time and can hence be used for high sampling rates. Hundreds of megahertz is common today. On the other hand, these converters are quite complex, they have limited word length and hence resolution (10 bits or less), they are quite expensive and often suffer from considerable power dissipation.

The block diagram of a simple 2 bit flash ADC is shown in Figure 2.12. The analog input is passed to a number of analog level comparators in parallel (i.e. a bank of fast operational amplifiers with high gain and low offset). If the analog input level U_{in} on the positive input of a comparator is greater than the level of the negative input, the output will be a digital 'one'. Else, the comparator outputs a digital 'zero'. Now, a reference voltage U_{ref} is fed to the voltage divider chain, thus obtaining a number of reference levels

$$U_k = \frac{k}{2^N} U_{ref} \tag{2.17}$$

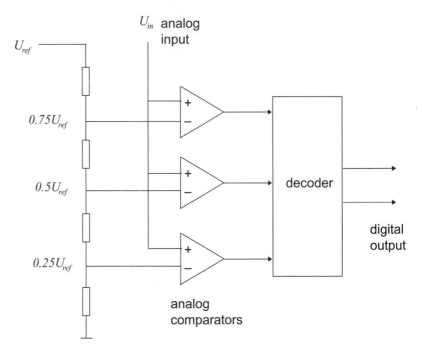

Figure 2.12 *An example 2 bit flash ADC*

where k is the quantization threshold number and N is the word length of the ADC. The analog input voltage will hence be compared to all possible quantization levels at the same time, rendering a 'thermometer' output of digital ones and zeros from the comparators. These ones and zeros are then used by a digital decoder circuit to generate digital parallel PCM data on the output of the ADC .

As pointed out above, this type of ADC is fast, but is hard to build for large word lengths. The resistors in the voltage divider chain have to be manufactured with high precision and the number of comparators and the complexity of the decoder circuit grow fast as the number of bits is increased.

2.3.3 Successive approximation A/D converters

These ADCs, also called **SAR** (successive approximation register) converters are the most common ones today. They are quite fast, but not as fast as flash converters. On the other hand they are easy to build and inexpensive, even for larger word lengths. The main parts of the ADC are: an analog comparator, a digital register, a DAC and some digital control logic (see Figure 2.13). Using the analog comparator, the unknown input voltage U_{in} is compared to a voltage U_{DAC} created by a DAC, being a part of the ADC. If the input voltage is greater than the voltage coming from the DAC, the output of the comparator is a logic 'one', else a logic 'zero'. The DAC is fed an input digital code from the register, which is in turn controlled by the control logic. Now, the principle of successive approximation works as follows.

Assume that the register contains all zeros to start with, hence, the output of the DAC is $U_{DAC} = 0$. Now, the control logic will start to toggle the MSB to a one, and the analog voltage coming from the DAC will be half of the maximum possible output voltage. The control logic circuitry samples the signal coming from the comparator. If this is a one, the control logic knows that the input voltage is still larger than the voltage coming from the DAC and the 'one' in the MSB will be left as is. If, on the other hand, the output of the comparator has turned zero, the output from the DAC is larger than the input voltage. Obviously, toggling the MSB to a one was just too much, and the bit is toggled back to zero. Now, the process is repeated for the second most significant bit and so on until all bits in the register have been toggled and set to a one or zero.

Hence, the SAR ADC always has a constant conversion time. It requires N approximation cycles, where N is the word length, i.e. the number of bits in the digital code. SAR-type converters of today may be used for sampling rates up to one megahertz.

An alternative way of looking at the converter is to see it as a DAC + register put in a control feedback loop. We try to 'tune' the register to match the analog input signal by observing the error signal from the comparator. Note that the DAC can of course be built in a variety of ways (see previous sections). Today, charge redistribution-based devices are quite common, since they are straightforward to implement using CMOS technology.

2.3.4 Counting A/D converters

An alternative, somewhat simpler ADC type is the **counting** ADC. The converter is mainly built in the same way as the SAR converter (Figure 2.13) but the control logic and register is simply a binary counter. The counter is reset to all zeros at the start of a conversion cycle, and is then incremented step by step. Hence, the output of the DAC is a staircase ramp function. The counting maintains until the comparator output switches to a zero and the counting is stopped.

The conversion time of this type of ADC depends on the input voltage, the higher the level, the longer the conversion time (counting is assumed to take place at a constant rate). The interesting thing about this converter is that the output signal of the comparator is a PWM representation of the analog input signal. Further, connecting an edge triggered monostable flip-flop to the comparator output, a PPM representation can also be obtained.

This type of converter is not very common. Using only a DAC and a comparator, the digital part can easily be implemented as software using a microcontroller. SAR type converters can of course also be implemented in the same way as well as 'tracking' type converters.

Tracking type converters can be seen as a special case of the generic counting converter. The tracking converter assumes that the change in the input signal level between consecutive samples is small. The counter is not restarted from zero at every conversion cycle, but starts from the previous state. If the output from the comparator is one, the counter is incremented until the comparator output toggles. If the output is zero when conversion

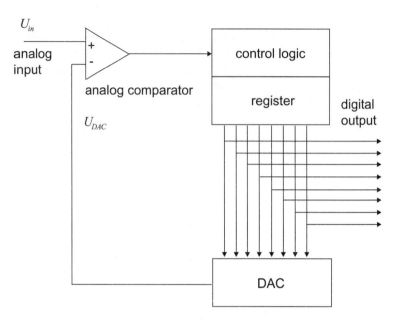

Figure 2.13 *An example Successive Approximation (SAR) ADC (simplified block diagram)*

is initiated, the counter is decremented until the output of the comparator toggles.

2.3.5 Integrating A/D converters

Integrating ADCs (sometimes also called counting converters) are often quite slow, but inexpensive and accurate. A common application is digital multimeters and similar equipment, where precision and cost are more important than speed.

There are many different variations of the integrating ADC, but the main idea is that the unknown analog voltage (or current) is fed to the input of an analog integrator with a well-known integration time constant $\tau = RC$. The slope of the ramp on the output of the integrator is measured by taking the time between the output level passing two or more fixed reference threshold levels. The time needed for the ramp to go from one threshold to the other is measured by starting and stopping a binary counter running at a constant speed. The output of the counter is hence a measure of the slope of the integrator output, which in turn is proportional to the analog input signal level. Since this type of ADC commonly has a quite long conversion time, i.e. integration time, the input signal is required to be stable or only slowly varying. On the other hand, the integration process will act as a low-pass filter, averaging the input signal and hence suppressing interference superimposed on the analog input signal to a certain extent.

Figure 2.14 shows a diagram of a simplified integrating ADC. The basic function (ramp method) works as follows. Initially, the input switch is

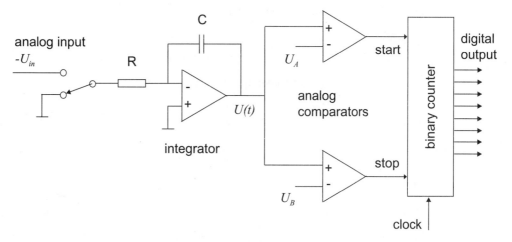

Figure 2.14 *A simplified integrating ADC*

connected to ground to reset the analog integrator. The switch then connects the unknown analog input voltage $-U_{in}$ to the input of the integrator. The output level of the integrator will then start to change as

$$U(t) = U_0 + U_{in}\frac{t - t_0}{RC} + \varepsilon\frac{t - t_0}{RC} \qquad (2.18)$$

where U_0 is the initial output signal from the integrator at time t_0 and ε is an error term representing the effect of offset voltage, leak currents and other shortcomings of a practical integrator. In practice, this term causes the output level of the integrator to slowly drift away from the starting level, even if the input voltage is held at zero.

Now, when the input signal is applied, the output level of the integrator starts to increase, and at time t_1 it is larger than the fixed threshold level U_A and a 'start' signal will be generated. The binary counter (that has been reset to zero) starts counting at a fixed rate, determined by the stable clock pulse frequency. The output level of the integrator continues to increase until, at time t_2, it is larger than the upper threshold U_B, and a 'stop' signal results. The binary counter stops and the binary PCM counter value now represents the time difference $t_2 - t_1$ and hence the analog input voltage

$$U(t_1) = U_A \quad \text{start signal at time } t_1 \quad U(t_2) = U_B \quad \text{stop signal at time } t_2$$

$$U_A = U(t_1) = U_0 + U_{in}\frac{t_1 - t_0}{RC} + \varepsilon\frac{t_1 - t_0}{RC}$$

$$= U_0 + (U_{in} + \varepsilon)\frac{t_1 - t_0}{RC} \qquad (2.19)$$

which can be rewritten as

$$t_1 - t_0 = (U_A - U_0)\frac{RC}{U_{in} + \varepsilon} \qquad (2.20)$$

and in a similar way, we obtain

$$t_2 - t_0 = (U_B - U_0)\frac{RC}{U_{in} + \varepsilon} \qquad (2.21)$$

Now, expressing the time difference measured by the counter we get

$$t_2 - t_1 = t_2 - t_0 - t_1 + t_0 = (U_B - U_A)\frac{RC}{U_{in} + \varepsilon} \qquad (2.22)$$

Rearranging equation (2.22) yields

$$U_{in} + \varepsilon = (U_B - U_A)\frac{RC}{t_2 - t_1} = \frac{K}{t_2 - t_1} \qquad (2.23)$$

As can be seen from equation (2.23) the unknown analog input voltage can easily be determined from the time difference recorded by the binary counter. Unfortunately, the error term will also be present. To obtain good precision in the ADC, we therefore need to design the circuitry carefully to reduce leak current etc. to a minimum. This is a problem associated with this type of ADC and is one of the reasons this basic conversion method is seldom used.

One way to eliminate the error term is to use a **dual slope** method. The term 'dual slope' refers to various methods in the literature, here only one method will be presented. Using this method, the timing measurement is performed in two phases. Phase 1 works as the simple ramp method described above. As soon as the output level of the integrator reaches the upper threshold, i.e. $U(t_2) = U_B$, phase 2 is initiated. During this phase, the polarity of the input signal is reversed and $+ U_{in}$ is fed to the input of the integrator. The output level of the integrator will now start to decrease and at time t_3, the lower threshold is reached again, i.e. $U(t_3) = U_A$, which completes the conversion process. We now have two time differences: $t_2 - t_1$ and $t_3 - t_2$ which can be used to express the analog input voltage with the error term eliminated. At time t_3 we have

$$U_A = U(t_3) = U(t_2) - U_{in}\frac{t_3 - t_2}{RC} + \varepsilon\frac{t_3 - t_2}{RC}$$

$$= U_B - (U_{in} - \varepsilon)\frac{t_3 - t_2}{RC} \qquad (2.24)$$

Rewriting equation (2.24) we get

$$-U_{in} + \varepsilon = -(U_B - U_A)\frac{RC}{t_3 - t_2} = -\frac{K}{t_3 - t_2} \qquad (2.25)$$

Subtracting term (2.25) from term (2.23) and dividing by two we obtain an expression for the analog input voltage without the error term

$$U_{in} = \frac{U_{in} + \varepsilon - (-U_{in} + \varepsilon)}{2}$$

$$= \frac{1}{2}\left((U_B - U_A)\frac{RC}{t_2 - t_1} + (U_B - U_A)\frac{RC}{t_3 - t_2}\right)$$

$$= \frac{K}{2}\left(\frac{1}{t_2 - t_1} + \frac{1}{t_3 - t_2}\right) = \frac{K}{2}\frac{t_3 - t_1}{(t_2 - t_1)(t_3 - t_2)} \tag{2.26}$$

An extension of the circuit in Figure 2.14 can be made in such a way the system is astable and continuously generates a square wave output having the period t_3-t_1. During t_2-t_1 the square wave is low and during t_3-t_2 it is high. A quite simple microcontroller can be used to calculate the value of the analog input voltage according to equation (2.26) above. An impressive resolution can be obtained using low-cost analog components.

2.3.6 Dither

Earlier, we discussed the use of dither techniques and oversampling to improve resolution of a DAC. Dither can be used in a similar way to improve the resolution of an ADC and hence reduce the effect of quantization distortion.

In the case of an ADC, analog noise with amplitude of typically 1/3 LSB is added to the input signal before quantization takes place. If the analog input signal level is between two quantization thresholds, the added noise will make the compound input signal cross the closest quantization level now and then. The closer to the threshold the original signal level, the more frequent the threshold will be crossed. Hence, there will be a stochastic modulation of the binary PCM code from the ADC containing additional information about the input signal. Averaging the PCM samples, resolution below the LSB can be achieved.

Dither is common in high quality digital audio systems (Pohlmann, 1989) nowadays, but has been used in video applications since 1950 and before that, to counteract gear backlash problems in early radar servo mechanisms.

2.3.7 Sigma–delta A/D converters

The **sigma–delta** ADC, sometimes also called **bitstream** ADC, utilizes the technique of oversampling, discussed earlier. The sigma–delta modulator was first introduced in 1962, but until recent developments in digital VLSI (very large scale integration) technology it was difficult to manufacture with high resolution and good noise characteristics at competitive prices.

One of the major advantages of the sigma–delta ADC using oversampling is that it is able to use digital filtering and relaxes the demands on the analog anti-aliasing filter. This also implies that about 90% of the die area is purely

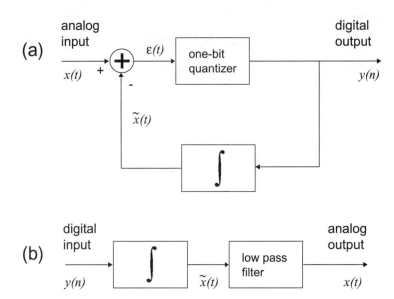

Figure 2.15 *A simplified delta modulator (a) and demodulator (b)*

digital, cutting production costs. Another advantage of using oversampling is that the quantization noise power is spread evenly over a larger frequency spectrum than the frequency band of interest. Hence, the quantization noise power in the signal band is lower than in traditional sampling based on the Nyquist criteria.

Now, let us take a look at a simple 1 bit sigma–delta ADC. The converter uses a method that was derived from the delta modulation technique. This is based on quantizing the **difference** between successive samples, rather than quantizing the absolute value of the samples themselves Figure 2.15 shows a **delta modulator** and demodulator. The modulator works as follows. From the analog input signal $x(t)$ a locally generated estimate $\tilde{x}(t)$ is subtracted. The difference $\varepsilon(t)$ between the two is fed to a 1 bit quantizer. In this simplified case, the quantizer may simply be the sign function, i.e. when $\varepsilon(t) \geqslant 0$, $y(n) = 1$, else $y(n) = 0$. The quantizer is working at the oversampling frequency, i.e. considerably faster than required by the signal bandwidth. Hence, the 1 bit digital output $y(n)$ can be interpreted as a kind of digital error signal:

$y(n) = 1$: estimated input signal level too small, increase level

$y(n) = 0$: estimated input signal level too large, decrease level

Now, the analog integrator situated in the feedback loop of the delta modulator is designed to function in exactly this way. Hence, if the analog input signal $x(t)$ is held at a constant level, the digital output $y(n)$ will (after convergence) be a symmetrical square wave (0 1 0 1 . . .), i.e. decrease, increase, decrease, increase . . . a kind of stable limit oscillation.

The delta demodulator is shown in the lower portion of Figure 2.15. The

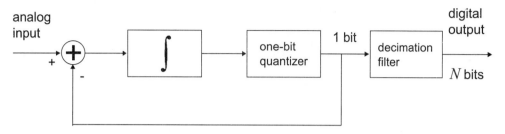

Figure 2.16 *A simplified, oversampled bitstream sigma–delta ADC*

function is straightforward. Using the digital 1 bit 'increase/decrease' signal, the estimated input level $\tilde{x}(t)$ can be created using an analog integrator of the same type as in the modulator. The output low-pass filter will suppress the ripple caused by the increase/decrease process.

Since integration is a linear process, the integrator in the demodulator can be moved to the input of the modulator. Hence, the demodulator will now only consist of the low-pass filter. We now have similar integrators on both inputs of the summation point in the modulator. For linearity reasons, these two integrators can be replaced by one integrator having $\varepsilon(t)$ connected to its input and the output connected to the input of the 1 bit quantizer. The delta modulator has now become a sigma–delta modulator. The name is derived from the summation point (sigma) followed by the delta modulator.

If we now combine the oversampled 1 bit sigma–delta modulator with a digital **decimation filter** (rate reduction filter) we obtain a basic sigma–delta ADC (see Figure 2.16). The task of the decimation filter is three-fold: to reduce the sampling frequency, to increase the word length from 1 bit to N bits and to reduce any noise pushed back into the frequency range of interest by the crude 1 bit modulator. A simple illustration of a decimation filter, decimating by a factor 5, would be an averaging process as shown in Table 2.2.

The decimation filter is commonly built using a decimator and a **comb filter** (Marven and Ewers, 1993; Lynn and Fuerst, 1994). The comb filter belongs to the class of frequency sampling filters and has the advantage of being easy to implement in silicon or in a DSP, using only additions and subtractions.

Table 2.2 *Decimation filter example*

Input:	0 0 1 1 0	0 1 1 1 0	1 1 1 0 1	1 0 1 0 1	0 0 0 1 1
Averaging process	3×0; 2×1	2×0; 3×1	1×0; 4×1	2×0; 3×1	3×0; 2×1
Output	0	1	1	1	0

3 Adaptive digital systems

3.1 Introduction   An adaptive signal processing system is a system which has the ability to change its processing behaviour in a way to maximize a given performance measure. An **adaptive system** is self-adjusting and is, by its nature, a time-varying and non-linear system. Hence, when using classical mathematical models and tools assuming linearity for the analysis of adaptive systems, care must be exercised.

A simple example of an adaptive system is the automatic gain control (AGC) in a radio receiver. When the input antenna signal is strong, the AGC circuitry reduces the amplification in the receiver to avoid distortion caused by saturation of the amplifying circuits. At weak input signals, the amplification is increased to make the signal readable.

Adaptive systems should, however, not be confused with pure feedback control systems. An electric heater controlled by a thermostat is, for instance, a feedback control system and not an adaptive system, since the control function is not changed (e.g. the thermostat always switches off the heater at 22°C).

The idea of self-adjusting systems (partly 'self-designing' systems) is not new. Building such systems using 'analog' signals and components is, however, very hard, except in some simple cases. The advent of VLSI digital signal processing and computing devices has made digital adaptive systems possible in practice.

In this chapter, we will discuss a class of adaptive digital systems based on closed-loop (feedback) adaptation. Some common systems and algorithms will be addressed. There are however, many variations possible (Widrow and Stearns, 1985). The theory of adaptive systems is on the borderline to optimization theory and neural network technology (addressed in Chapter 4).

3.1.1 System structure

Figure 3.1 shows a generic adaptive signal processing system. The system consists of three parts: the processor, the performance function and the adaptation algorithm.

The **processor** is the part of the system that is responsible for the actual processing of the input signal x, thus generating the output signal y. The processor can, for instance, be a digital FIR filter.

The **performance function** takes the signals x and y as inputs as well as 'other data' d, that may affect the performance of the entire system from x to y. The performance function is a quality measure of the adaptive system. In optimization theory, this function corresponds to the 'objective function'

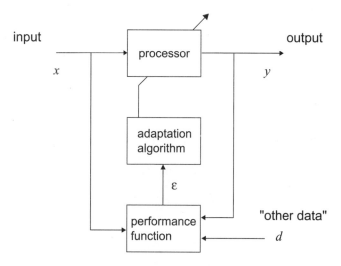

Figure 3.1 *A generic closed loop adaptive system*

and in control theory it corresponds to the 'cost function'. The output ε from the performance function is a 'quality signal' illustrating the processor at its present state and indicating whether it is performing well, taking into account the input signals, output signals and other relevant parameters.

The quality signal ε is finally fed to the **adaptation algorithm**. The task of the adaptation algorithm is to change the parameters of the processor in such a way that performance is maximized. In the example FIR filter, this would imply changing the tap weights.

Closed-loop adaptation has the advantage of being usable in many situations where no analytic synthesis procedure either exists or is known and/or when the characteristics of the input signal varies considerably. Further, in cases where physical system component values are variable or inaccurately known, closed-loop adaptation will find the best choice of parameters. In the event of partial system failure, the adaptation mechanism may even be able to readjust the processor in a way that the system will still achieve an acceptable perform-ance. System reliability can often be improved using performance feedback.

There are, however, some inherent problems as well. The processor model and the way the adjustable parameters of the processor model is chosen affects the possibilities of building a good adaptive system. We must be sure that the desired response of the processor can really be achieved by adjusting the parameters within the allowed ranges. Using too complicated a processor model would on the other hand make analytic analysis of the system cumber-some or even impossible.

The performance function also has to be chosen carefully. If the perfor-mance function does not have a unique extremum, the behaviour of the adaptation process may be uncertain. Further, in many cases it is desirable (but not necessary) that the performance function is differentiable. Choosing 'other data' carefully may also affect the usefulness of the performance func-tion of the system.

Finally, the adaptation algorithm must be able to adjust the parameters of the processor in a way to improve the performance as fast as possible and to eventually converge to the optimum solution (in the sense of the performance function). Too 'fast' an adaptation algorithm may, however, result in instability or undesired oscillatory behaviour. These problems are well known from optimization and control theory.

3.2 The processor and the performance function

In this section, a common processor model, the **adaptive linear combiner**, and a common performance function, the **mean square error** will be presented. Throughout this text, vector notation will be used, where all vectors are assumed to be column vectors and are usually indicated as a transpose of a row vector, unless stated otherwise.

3.2.1 The adaptive linear combiner

One of the most common processor models is the so-called adaptive linear combiner (Widrow and Stearns, 1985), shown in Figure 3.2. In the most general form, the combiner is assumed to have $L+1$ inputs denoted

$$X_k = \begin{bmatrix} x_{0k} & x_{1k} & \cdots & x_{Lk} \end{bmatrix}^T \tag{3.1}$$

where the subscript k is used as the time index. Similarly, the gains or weights of the different branches are denoted

$$W_k = \begin{bmatrix} w_{0k} & w_{1k} & \cdots & w_{Lk} \end{bmatrix}^T \tag{3.2}$$

Since we are now dealing with an adaptive system, the weights will also vary in time and hence have a time subscript k. The output y of the adaptive multiple input linear combiner can be expressed as

$$y_k = \sum_{l=0}^{L} w_{lk} x_{lk} \tag{3.3}$$

As can be seen from equation (3.3) this is nothing but a dot product between the input vector and the weight vector, hence

$$y_k = X_k^T W_k = W_k^T X_k \tag{3.4}$$

The adaptive linear combiner can also take another form working in the temporal domain, with only a single input (see Figure 3.3). This is the general form of the adaptive linear combiner shown in Figure 3.2, extended with a delay line. In this case the output is given by the convolution sum

$$y_k = \sum_{l=0}^{L} w_{lk} x_{k-l} \tag{3.5}$$

which is the expression for the well-known FIR filter or transversal filter. Now, if the input vector is defined as

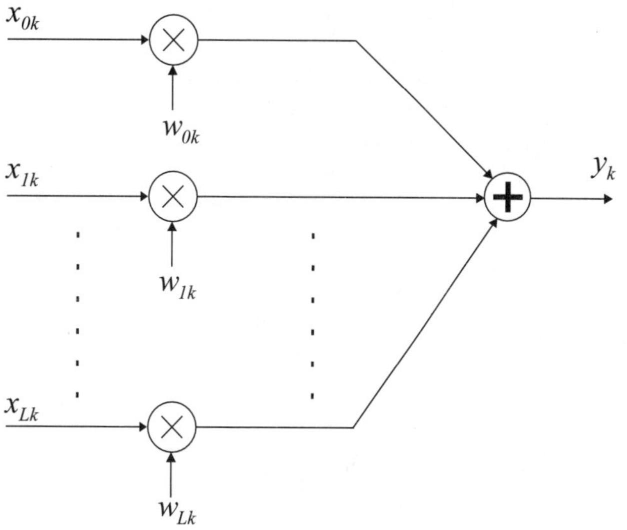

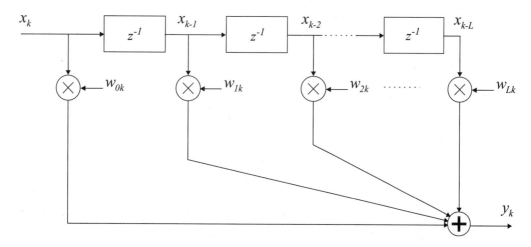

Figure 3.2 *The general form of an adaptive linear combiner with multiple inputs*

$$X_k = \begin{bmatrix} x_k & x_{k-1} & \cdots & x_{k-L} \end{bmatrix}^T \tag{3.6}$$

equation (3.4) still holds. It is possible to use other types of structure, e.g. IIR filters or lattice filters as processors in an adaptive system. The adaptive linear combiner is, however, by far the most common, since it is quite straightforward to analyse and design. It is much harder to find good adaptation algorithms for IIR filters, since the weights are not permitted to vary arbitrarily because instability and oscillations may occur.

Figure 3.3 *A temporal version of an adaptive linear combiner with a single input – an adaptive transversal filter (FIR filter or tapped delay line filter)*

3.2.2 The performance function

The performance function can be designed in a variety of ways. In this text, we will concentrate on the quite common **mean square error (MSE)** measure. To be able to derive the following expressions, we will assume that 'other data' d_k will be the desired output from our system, sometimes called the '**training signal**' or '**desired response**' (see Figure 3.4). One could argue: why employ an adaptive system at all if the desired response is known in advance? However, presently we shall assume the availability of such a signal. Later, we will discuss its derivation in more detail. Considerable ingenuity is however needed in most cases to find suitable 'training signals'.

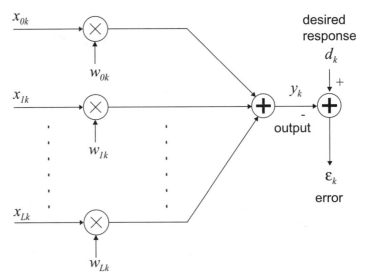

Figure 3.4 *Adaptive linear combiner with desired response ('other data') and error signals*

In this case, the performance function will be based on the error signal ε_k

$$\varepsilon_k = d_k - y_k = d_k - X_k^T W = d_k - W^T X_k \qquad (3.7)$$

In this discussion, we assume that the adaptation process is slow compared to the variations in time of the input signal. Hence, the time subscript on the weight vector has been dropped. From equation (3.7) we will now derive the performance function as the MSE. We then have to minimize the MSE (the power of the error signal) to maximize the performance of the corresponding adaptive system. Now, take the square of equation (3.7)

$$\varepsilon_k^2 = (d_k - W^T X_k)(d_k - X_k^T W) = d_k^2 + W^T X_k X_k^T W - 2d_k X_k^T W \qquad (3.8)$$

Assume that ε_k, d_k and X_k are statistically stationary (Papoulis, 1985), and take the expected value of equation (3.8) over k to obtain the MSE

$$\xi = \mathrm{E}[\varepsilon^2] = \mathrm{E}[d_k^2] + \boldsymbol{W}^T\mathrm{E}[\boldsymbol{X}_k\boldsymbol{X}_k^T]\boldsymbol{W} - 2\,\mathrm{E}[d_k\boldsymbol{X}_k^T]\boldsymbol{W} \tag{3.9}$$

$$= \mathrm{E}[d_k^2] + \boldsymbol{W}^T\boldsymbol{R}\boldsymbol{W} - 2\boldsymbol{P}^T\boldsymbol{W}$$

where the matrix $\boldsymbol{R}$ is the square input **correlation matrix** (Papoulis, 1985). The main diagonal terms are the mean squares of the input components, i.e. the variance and the power of the signals. The matrix is symmetric. If the single input form of the linear combiner is used this matrix constitutes the **auto-correlation matrix** (Papoulis, 1985) of the input signal. In the latter case, all elements on the diagonal will be equal.

The matrix $\boldsymbol{R}$ appears as

$$\boldsymbol{R} = \mathrm{E}\,[\boldsymbol{X}_k\boldsymbol{X}_k^T] = \mathrm{E}\begin{bmatrix} x_{0k}^2 & x_{0k}x_{1k} & \cdots & x_{0k}x_{Lk} \\ x_{1k}x_{0k} & x_{1k}^2 & \cdots & x_{1k}x_{Lk} \\ \vdots & \vdots & & \vdots \\ x_{Lk}x_{0k} & x_{Lk}x_{1k} & & x_{Lk}^2 \end{bmatrix} \tag{3.10}$$

The vector $\boldsymbol{P}$ is the **cross-correlation** (Papoulis, 1985) between the desired response and the input components

$$\boldsymbol{P} = \mathrm{E}\big[d_kx_{0k}\;\; d_kx_{1k}\;\; \cdots\;\; d_kx_{Lk}\big]^T \tag{3.11}$$

Since the signals d_k and x_{ik} are generally not statistically independent, the expected value of the product $d_k\boldsymbol{X}_k^T$ **cannot** be rewritten as a product of expected values. If d_k and all x_{ik} are indeed statistically independent, they will also be uncorrelated and $\boldsymbol{P} = \boldsymbol{0}$. It can easily be seen from equation (3.9) that in this case, minimum MSE would occur when setting $\boldsymbol{W} = \boldsymbol{0}$, i.e. when setting all weights equal to zero, or in other words switching the processor off completely. It does not matter what the adaptive system tries to do using the input signals, MSE cannot be reduced using any input signal or combination thereof. In this case, we have probably made a poor choice of d_k or $\boldsymbol{X}_k$, or the processor model is not relevant.

From equation (3.9) we can see that the MSE is a quadratic function of the components of the weight vector. This implies that the surface of the performance function will be bowl shaped and form a hyperparaboloid in the general case, which only has **one global minimum**. Figure 3.5 shows an example two-dimensional (paraboloid) quadratic performance surface. The minimum MSE point, i.e. the optimum combination of weights, can be found at the bottom of the bowl. This set of weights is sometimes called the Wiener weight vector $\boldsymbol{W}^*$. The optimum solution, i.e. the Wiener weight vector can be found by finding the point where the gradient of the performance function is zero. Differentiating equation (3.9) we obtain the gradient

$$\nabla(\xi) = \frac{\partial\xi}{\partial\boldsymbol{W}} = \left[\frac{\partial\xi}{\partial w_0}\;\; \frac{\partial\xi}{\partial w_1}\;\; \cdots\;\; \frac{\partial\xi}{\partial w_L}\right]^T = 2\boldsymbol{R}\boldsymbol{W} - 2\boldsymbol{P} \tag{3.12}$$

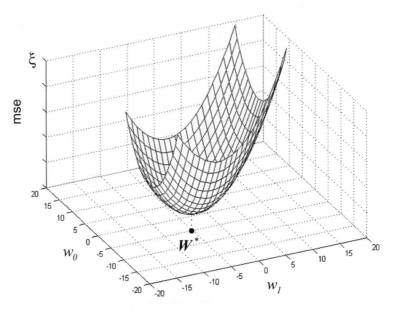

Figure 3.5 *Two-dimensional paraboloid performance surface. The MSE as a function of two weight components w_0 and w_1. Minimum error, i.e. optimum, is found at the bottom of the bowl, i.e. at W^*, the Wiener weight vector*

The optimum weight vector is found where the gradient is zero, hence

$$\nabla(\xi) = \mathbf{0} = 2RW^* - 2P \tag{3.13}$$

Assuming that R is non-singular, the Wiener–Hopf equation in matrix form is

$$W^* = R^{-1}P \tag{3.14}$$

From this we realize that adaptation (optimization), i.e. finding the optimum weight vector, is an iterative method of finding the inverse (or pseudoinverse) of the input correlation matrix.

The minimum mean square error (at optimum) is now obtained by substituting W^* from equation (3.14) for W in equation (3.9) and using the symmetry of the input correlation matrix, i.e. $R^T = R$ and $(R^{-1})^T = R^{-1}$

$$\begin{aligned}
\xi_{min} &= \mathrm{E}\!\left[d_k^2\right] + W^{*T}RW^* - 2P^TW^* \\
&= \mathrm{E}\!\left[d_k^2\right] + (R^{-1}P)^T R R^{-1}P - 2P^TR^{-1}P \\
&= \mathrm{E}\!\left[d_k^2\right] - P^TR^{-1}P = \mathrm{E}\!\left[d_k^2\right] - P^TW^* \tag{3.15}
\end{aligned}$$

Finally, a useful and important statistical condition exists between the error signal ε_k and the components of the input signal vector X_k when $W = W^*$. Multiplying equation (3.7) by X_k from the left, we obtain

$$\varepsilon_k X_k = d_k X_k - y_k X_k = d_k X_k - X_k X_k^T W \tag{3.16}$$

Taking the expected value of equation (3.16) yields

$$\mathrm{E}\left[\varepsilon_k X_k\right] = \mathrm{E}\left[d_k X_k\right] - \mathrm{E}\left[X_k X_k^T\right]W = P - RW \tag{3.17}$$

Inserting equation (3.14) into equation (3.17) gives

$$\mathrm{E}\left[\varepsilon_k X_k\right]_{W=W^*} = P - RR^{-1}P = P - P = 0 \tag{3.18}$$

This result corresponds to the result of the Wiener filter theory. When the impulse response (weights) of a filter is optimized, the error signal is uncorrelated (orthogonal to) the input signals.

3.3 Adaptation algorithms

From the above we have realized that the mean square error performance surface for the linear combiner is a quadratic function of the weights (input signal and desired response statistically stationary). The task of the adaptation algorithm is to locate the optimum setting W^* of the weights, hence obtaining the best performance from the adaptive system. Since the parameters of the performance surface may be unknown and analytical descriptions are not available, the adaptation process is an iterative process, searching for the optimum point.

Many different adaptation schemes have been proposed. In this text a few common generic types will be discussed. The ideal adaptation algorithm converges quickly but without oscillatory behaviour, to the optimum solution. Once settled at the optimum, the ideal algorithm should track the solution if the shape of the performance surface changes over time. The adaptation algorithm should preferably also be easy to implement, and not be excessively demanding with regard to the computations.

The shape of the **learning curve**, i.e. a plot of the MSE ξ_i as a function of the iteration number i, is in many cases a good source of information about the properties of an adaptation algorithm. Note! For every iteration period i there is commonly a large number of sample instants k, since ξ is the average over k (time).

3.3.1 The method of steepest descent

The idea of the steepest descent method is as follows: starting in an arbitrary point W_0 of the performance surface, estimate the gradient and change the weights in the negative direction of the gradient ∇_0. In this way, the weight vector point will proceed 'downhill', until reaching the bottom of the performance surface 'bowl' and the optimum point. The steepest descent algorithm can be expressed as

$$W_{i+1} = W_i + \mu(-\nabla_i) \tag{3.19}$$

where μ is the step size at each iteration. The larger the step size, the faster the convergence. An excessively large step size would, on the other hand, cause instability and oscillatory problems in the solution. This problem is common to all closed-loop control systems. Finding a good compromise on μ may therefore be a problem.

Another problem is obtaining the gradient ∇_i. Since the gradient is not known in advance in the general case, it has to be estimated by perturbing the weights one by one by a small increment $\pm\delta$ and measuring the MSE ξ. Hence, the gradient is expressed as

$$\nabla(\xi) = \frac{\partial \xi}{\partial W} = \left[\frac{\partial \xi}{\partial W} \frac{\partial \xi}{\partial w_1} \cdots \frac{\partial \xi}{\partial w_L} \right]^T \tag{3.20}$$

where the respective derivatives can be estimated by

$$\frac{\partial \xi}{\partial w_n} \approx \frac{\xi(w_n + \delta) - \xi(w_n - \delta)}{2\delta} \tag{3.21}$$

One has to remember that the performance surface is a 'noisy' surface in the general case. To obtain a MSE with a small variance, many samples are required. There will be two perturbation points for each dimension of the weight vector; $2L$ averaged measurements, each consisting of a number of samples are needed. There is of course a tradeoff. The larger the number of averaged points, in other words the longer the estimation time, the less noise there will be in the gradient estimate. A noisy gradient estimate will result in an erratic adaptation process.

Disregarding the noise, it should also be noted that the steepest descent 'path' is not necessarily the straight route to the optimum point in the general case. An example is shown in Figure 3.6a. Looking down into the bowl-shaped performance function of Figure 3.5, the movement of the weight vector during the iterations is shown. Starting in a random point, the change of the weight vector is always in the direction of the negative gradient (steepest descent).

3.3.2 Newton's method

Newton's method may be seen as an improved version of the steepest descent method discussed above. In this method, the steepest descent route is not used, but the weight vector moves 'directly' towards the optimum point. This is achieved by adding information about the shape of the surface to the iterative adaptation algorithm

$$W_{i+1} = W_i + \mu R^{-1}(-\nabla_i) \tag{3.22}$$

The extra information is introduced into equation (3.22) by including R^{-1}, the inverse of the input correlation matrix. When the gradient is multiplied by this matrix, the resulting move of the weight vector will not be in the negative direction of the gradient, but rather in the direction towards the optimum point.

Due to the additional matrix multiplication, Newton's method requires more computational power than the method of steepest descent. In addition, knowledge of the inverse of the input correlation matrix is required, which in turn requires averaging of a number of samples.

The advantage of this method is that it often finds the optimum point in very few iterations. It is easy to show that if we have a perfect knowledge of the gradient and the inverse of the input correlation matrix Newton's method will find the optimum solution in only one iteration. Setting $\mu = 0.5$ in equation (3.22) and inserting equation (3.13) we obtain

$$
\begin{aligned}
W_{i+1} &= W_i - 0.5R^{-1}\nabla_i \\
&= W_i - 0.5R^{-1}(2RW_i - 2P) \\
&= R^{-1}P = W^* \quad\quad\quad (3.23)
\end{aligned}
$$

where the last equality is given by equation (3.14). Hence, it does not matter where we start on the performance surface, if the gradient and input signal inverse correlation matrix are known, the optimum point will be reached in

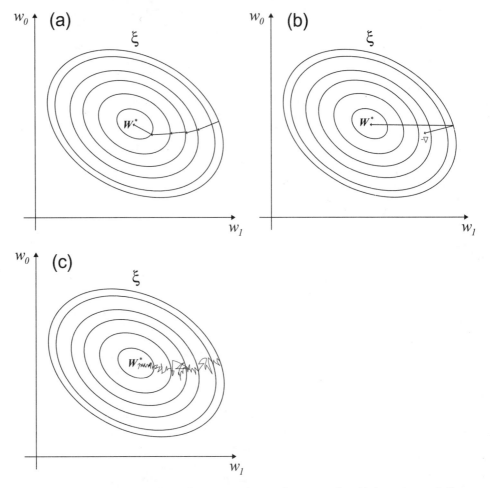

Figure 3.6 *Looking down into the performance 'bowl' (see Figure 3.5), iteration of the weight vector towards optimum at* W^**. (a) steepest descent. (b) Newton's method. (c) LMS*

only **one** iteration (see Figure 3.6b). In practice, this is not possible in the general case, since the gradient estimate and the elements of the input correlation matrix will be noisy.

3.3.3 The LMS algorithm

The methods presented above, i.e. the method of steepest descent and Newton's method, both require an estimation of the gradient at each iteration. In this section, the **least mean square** (LMS) algorithm will be discussed. The LMS algorithm uses a special estimate of the gradient that is valid for the adaptive linear combiner. Thus, the LMS algorithm is more restricted in its use than the other methods.

On the other hand, the LMS algorithm is important because of its ease of computation and because it does not require the repetitive gradient estimation. The LMS algorithm is quite often the best choice for many different adaptive signal processing applications. Starting from equation (3.7) we recall that

$$\varepsilon_k = d_k - y_k = d_k - X_k^T W \tag{3.24}$$

In the previous methods, we would estimate the gradient using $\xi = E[\varepsilon_k^2]$, but the LMS algorithm uses ε_k^2 itself as an estimate of ξ. This is the main point of the LMS algorithm. At each iteration in the adaptation process, we have a gradient estimate of the form

$$\hat{\nabla}_k = \begin{bmatrix} \partial \varepsilon_k^2/\partial w_0 \\ \partial \varepsilon_k^2/\partial w_1 \\ \vdots \\ \partial \varepsilon_k^2/\partial w_L \end{bmatrix} = 2\varepsilon_k \begin{bmatrix} \partial \varepsilon_k/\partial w_0 \\ \partial \varepsilon_k/\partial w_1 \\ \vdots \\ \partial \varepsilon_k/\partial w_L \end{bmatrix} = -2\varepsilon_k X_k \tag{3.25}$$

where the derivatives of ε_k follow from equation (3.24). Using this simplified gradient estimate, we can now formulate a steepest descent type algorithm:

$$W_{k+1} = W_k - \mu \hat{\nabla}_k = W_k + 2\mu \varepsilon_k X_k \tag{3.26}$$

This is the LMS algorithm. The gain constant μ governs the speed and stability of the adaptation process. The weight changes at each iteration are based on imperfect, noisy gradient estimates, which implies that the adaptation process does not follow the true line of steepest descent on the performance surface (see Figure 3.6c). The noise in the gradient estimate is however attenuated with time by the adaptation process, which acts as a low-pass filter.

The LMS algorithm is elegant in its simplicity and efficiency. It can be implemented without squaring, averaging or differentiation. Each component of the gradient estimate is obtained from a single data sample input, no perturbations are needed.

The LMS algorithm takes advantage of prior information regarding the quadratic shape of the performance surface. This gives the LMS algorithm considerable advantage over the previous adaptation algorithms when it comes to adaptation time. The LMS algorithm converges to a solution much faster than the steepest descent method, particularly when the number of weights is large.

3.4 Applications

In this section some example applications of adaptive systems will be briefly presented. The presentation is somewhat simplified and some practical details have been omitted, thus simplifying the understanding of the system's main feature.

3.4.1 Adaptive interference cancelling

Adaptive interference cancelling devices can be used in many situations where a desired signal is disturbed by additive background noise. This could be, for instance, the signal from the headset microphone used by the pilot in an aircraft. In this case, the speech signal is disturbed by the background noise created by motors, propellers and air flow. It could also be the weak signal coming from electrocardiographic electrodes, being disturbed by 50 or 60 Hz interference caused by nearby power lines and appliances.

The idea behind the interference cancelling technique is quite simple. Assume that our desired signal s is disturbed by additive background noise n_0. The available signal is

$$x = s + n_0 \tag{3.27}$$

If we could only obtain the noise signal n_0, it could easily be subtracted from the available signal and the noise would be cancelled completely. In most cases, the noise signal n_0 is not available, but if a correlated noise signal n_1 can be obtained, n_0 may be created by filtering n_1. Since the filter function needed is not known in most cases, this is a perfect job for an adaptive filter. An adaptive noise cancelling circuit is shown in Figure 3.7. As can be seen from the figure, the output signal is fed back as an error signal

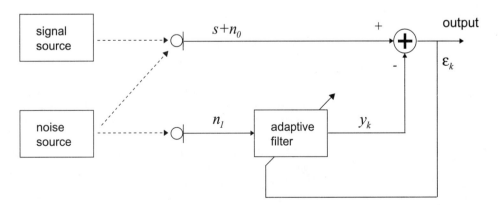

Figure 3.7 *An adaptive noise-cancelling circuit, using an adaptive filter*

ε to the adaptation algorithm of the adaptive filter. Assume that s, n_0, n_1 and y are statistically stationary and have zero means. Further, assume that s is uncorrelated with n_0 and n_1, and that n_1 is correlated with n_0, then

$$\varepsilon = s + n_0 - y \tag{3.28}$$

Squaring equation (3.28) gives

$$\varepsilon^2 = s^2 + (n_0 - y)^2 + 2s(n_0 - y) \tag{3.29}$$

Taking expectations of both sides and making use of the fact that s is uncorrelated with n_0 and y, we get

$$\begin{aligned}
E[\varepsilon^2] &= E[s^2] + E[(n_0 - y)^2] + 2E[s(n_0 - y)] \\
&= E[s^2] + E[(n_0 - y)^2]
\end{aligned} \tag{3.30}$$

From this equation it can seen that adapting the filter by minimizing the mean square error $E[\varepsilon^2]$ i.e. the output signal power, is done by minimizing the term $E[(n_0 - y)^2]$, since the input signal power $E[s^2]$ cannot be affected by the filter. Minimizing the term $E[(n_0 - y)^2]$ implies that the output y of the filter is a best least-squares estimate of unknown noise term n_0, which is needed to cancel the noise in the input signal. Hence, from equation (3.28) we realize that minimizing $E[(n_0 - y)^2]$ implies that $E[(\varepsilon_0 - s)^2]$ and the noise power of the output signal are also minimized.

Let us pursue the previous example of cancelling the background noise n_0 picked up by the pilot's headset microphone. In a practical situation, we would set up a second microphone on the flight deck in such a way that it only picks up the background noise n_1. Using a device like that in Figure 3.7, the background noise n_0 disturbing the speech signal s could be suppressed. Recently, the same technique has been adopted for cellular telephones as well.

Other areas where adaptive interference cancelling systems are used are echo cancellation (Widrow and Stearns, 1985) in telephone lines where echo may occur as a result of unbalanced hybrids and interference suppression in radio systems.

The performance of the adaptive noise cancelling system depends on three factors. First, there must be a fair correlation between the noise n_1 picked up by the 'reference microphone' and the disturbing background noise n_0. If they are completely uncorrelated, the adaptive filter will set all its weights to zero (i.e. 'switch off') and will have no effect (no benefit or damage). Second, the filter must be able to adapt to such a filter function so that n_0 can be well created from n_1. This implies that the processor model used in the filter is relevant and that the adaptation algorithm is satisfactory. Note that we have not assumed that the adaptive filter necessarily converges to a linear filter. Third, it is preferable that there be no correlation between the signal n_1 picked up by the reference microphone and the desired signal s. If there is a correlation, the system will also try to cancel the desired signal, that is, reduce $E[s^2]$. Hence, we have to choose the reference noise signal carefully.

3.4.2 Equalizers

Equalizers or adaptive inverse filters are frequently used in many telecommunication applications where a transmitted signal is being distorted by some filtering function inherent in the transmission process. One example is the limited bandwidth of a telephone line (filter effect) that will tend to distort high-speed data transmissions. Now, if the transfer function $H(\omega)$ from transmitter to receiver of the telephone line is known, we can build a filter having the inverse transfer function $G(\omega) = H^{-1}$ a so-called **inverse filter** or **zero-forcing filter**. If the received signal is fed to the inverse filter, the filtering effect of the telephone line can be neutralized, such that

$$H(\omega)\,G(\omega) = H(\omega)\,H^{-1}(\omega) = 1 \qquad (3.31)$$

The transfer function of a communication channel (e.g. telephone line) can be modelled as a **channel filter** $H(\omega)$. Usually, the transfer function is not known in advance and/or varies over time. This implies that a good fixed inverse filter cannot be designed, but an adaptive inverse filter would be usable. This is what an equalizer is, an adaptive inverse filter that tries to neutralize the effect of the channel filter at all times in some optimum way. Today's high-speed telephone modems would not be possible without the use of equalizers.

Figure 3.8 shows a block diagram of a transmission system with a channel filter having transfer function $H(z)$ and an equalizer made up of an adaptive filter used as inverse filter having transfer function $G(z)$. The signal s_k entering the transmission channel is distorted by the transfer function of the channel filter, but on top of this additive noise n_k is disturbing the transmission, hence the received signal is

$$r_k = s_k * h(k) + n_k \qquad (3.32)$$

In the **channel model** above, * denotes convolution and $h(k)$ is the impulse response corresponding to the channel filter transfer function $H(z)$. Now, to make the adaptive filter converge resulting in a good inverse transfer func-

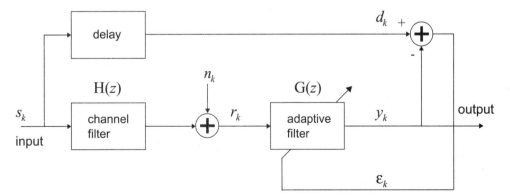

Figure 3.8 *A transmission situation with channel filter, additive noise and an equalizer implemented by means of an adaptive filter*

tion, a desired signal d_k is needed. If the transmitted signal s_k is used as a desired signal, the adaptive filter will converge in a way to minimize the mean square error

$$E\left[\varepsilon_k^2\right] = E\left[(s_k - y_k)^2\right] \qquad (3.33)$$

implying that the output y_k will resemble the input s_k as closely as possible in the least-square sense, and that the adaptive filter has converged to a good inverse $G(z)$ of the channel filter transfer function $H(z)$. The inverse filter will eventually also track variations in the channel filter.

Finding a desired signal d_k is not always entirely easy. In the present example, we have used the transmitted signal s_k, which is not in the general case known at the receiving site. A common method is to transmit known 'training signals' now and then, to be used for adaptation of the inverse filter. There are also other ways of defining the error signal. For instance, in a digital transmission system (where s_k is 'digital'), the error signal can be defined as

$$\varepsilon_k = \hat{s}_k - y_k \qquad (3.34)$$

where $\hat{s}_k$ is the digital **estimate** of the transmitted (digital) signal s_k and $\hat{s}_k$ is the output of the non-linear detector using the analog signal y_k as input. In this case, the detector simply consists of an analog comparator. Hence, the estimate of the transmitted signal is used as the desired signal instead of the transmitted signal itself (see Figure 3.9). As the equalizer converges, the analog signal y_k will be forced stronger and stronger when $\hat{s}_k = 1$ and weaker and weaker when $\hat{s}_k = 0$.

There are two problems associated with equalizers. First, since the zeros in the channel filter transfer function will be reflected as poles in the inverse filter, we may not be able to find a stable inverse filter for all channel filters. Second, assume that there is a deep notch in the channel filter function at a specific frequency. This notch will be counteracted by a sharp peak, having a high gain at the same frequency in the transfer function of the inverse filter. The additive noise, possibly being stronger than the desired signal at

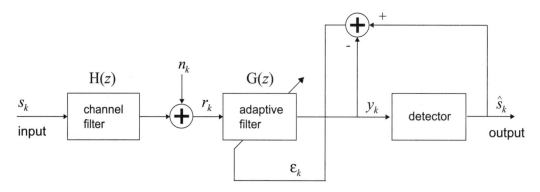

Figure 3.9 *An equalizer using the estimated transmitted signal as the desired signal*

this specific frequency, will then be heavily amplified. This may cause a very poor signal-to-noise ratio, rendering the equalized output signal unreadable.

Equalizers are sometimes also used in high performance audio systems to counteract the transfer function caused by room acoustics. In this situation however, the transmitted signal is 'precompensated' before being sent to the loudspeakers. Hence, the order of the inverse filter and the channel filter is reversed compared to Figure 3.8.

For equalizing telephone lines and room acoustics, equalizers using a simple adaptive linear inverse filter will often do well. When it comes to equalizers for data transmission over radio links, more complex equalizers are often required to achieve acceptable performance. Different types of **decision-feedback equalizers** are common in digital radio transmission systems (Proakis, 1989; Ahlin and Zander, 1998).

In a radio link (Ahlin and Zander, 1998) the signal travelling from the transmitter to the receiver antenna will be propagated over many different paths simultaneously, all the paths having different length. The multitude of received signal components compounded in the receiver antenna will hence be delayed different amounts of time, which will give the components different phase shift. This in turn results in a filtering effect of the transmitted signal (the channel filter). The phenomenon is denoted **multi-path propagation** (Ahlin and Zander, 1998), causing **frequency selective fading** (Ahlin and Zander, 1998). Due to the transfer function of the channel filter, the short data bits will be 'smeared' out, causing interference in the following data bit time slots. This is called **intersymbol interference ISI** (Ahlin and Zander, 1998), and can be counteracted by including an equalizer in the receiver.

Figure 3.10 shows an example decision-feedback equalizer. This is mainly the same equalizer as in Figure 3.9, but extended with a feedback loop consisting of another adaptive filter. The latter filter 'remembers' the past detected digital symbols and counteracts the residuals of the intersymbol

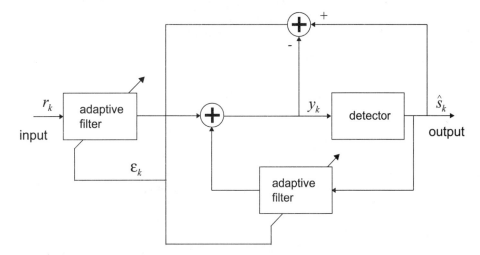

Figure 3.10 *A decision-feedback equalizer (channel filter and additive noise not shown)*

interference. Both filters are adapted synchronously using a common adaptation algorithm.

This type of non-linear equalizer works well for combating intersymbol interference present in highly time-dispersive radio channels. If we assume that both filters are adaptive FIR filters, the output of the equalizer will be

$$y_k = \sum_{l=0}^{L} w_{lk} r_{k-1} + \sum_{j=0}^{J} v_{jk} \hat{s}_{j-k} \tag{3.35}$$

where w_{lk} are the weights of the first filter and v_{jk} the weight of the feedback filter at time instant k.

3.4.3 Adaptive beamforming

So far we have only discussed the use of adaptive systems in the time and frequency domains, in the single input processor model, according to equations (3.5) and (3.6). In this section applications using the more general **multiple-input** linear combiner model and signal conditioning in the **spatial domain** will be addressed.

The underlying signal processing problem is common to the reception of many different types of signals, such as electromagnetic, acoustic or seismic. In this signal processing-oriented discussion, the difference between these situations is simply the choice of sensors: radio antennas, microphones, hydrophones, seismometers, geophones etc. Without sacrificing generality, we will use the reception of electromagnetic signals by radio antennas as an example in the following discussion.

Assume the we are interested in receiving a fairly weak signal in the presence of a strong interfering signal. The desired signal and the interfering signal originate from different locations and at the receiving site, the signals have different angles of arrival. A good way of dealing with this problem is to design a directive antenna, an antenna having a maximum sensitivity in one direction (the 'beam' or 'lobe') and a minimum sensitivity in other directions. The antenna is directed in such a way that the maximum sensitivity direction coincides with the direction of the location of the desired signal, and minimum sensitivity ('notch') angles in the direction of the interferer (see Figure 3.11).

There are however some problems. The physical size of a directive antenna depends on the wavelength of the signal. Hence, for low frequency signals, the wavelength may be so long that the resulting directive antenna would be huge and not possible to build in practice. Another problem occurs if the desired signal source or the interferer or both are mobile, i.e. (fast) moving. In such a case, the directive antenna may need to be redirected continuously (for instance radar antennas). If the directive antenna structure needed to be moved (quickly) this may create challenging mechanical problems.

An alternative way of building a directive antenna is to use a number of omnidirectional, fixed antennas mounted in an **antenna array**. The output of the antennas are fed to a signal processing device, and the resulting directivity pattern of all the antennas in unison is created 'electronically' by the

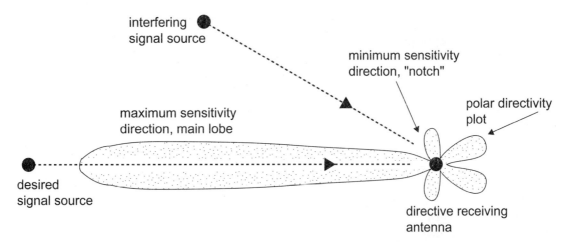

Figure 3.11 *Enhancing the desired signal and attenuating the interference using a directive antenna (the directivity plot is a polar diagram showing the relative sensitivity of the antenna as a function of the angle)*

signal processing device. Hence, the shape and direction of the lobes and notches can be changed quickly, without having to move any part of the physical antenna system. The antenna system may however still be large for low frequencies, but for many cases easier to build in practice than in the previous case.

The drawback of the 'electronic' directive antenna is the cost and complexity. The advent of fast and inexpensive digital signal processing components has however made this approach known as **beamforming** (Widrow and Stearns, 1985) more attractive during the past years. The steady increase in processing speed of digital signal processing devices has also made it possible to use this technology for high-frequency signal applications a task impossible a few of years back.

The following simplified example (see Figure 3.12) illustrates the use of a multiple-input, adaptive linear combiner in a beamforming 'electronic' directive antenna system. Assume that we are using two omnidirectional antennas A_1 and A_2. The distance between the two antennas is l. We are trying to receive a weak desired narrowband signal $s \cos(\omega t)$, but unfortunately we are disturbed by a strong narrowband, jamming signal of the same frequency $u \cos(\omega t)$. The desired signal is coming from a direction perpendicular to the normal of the antenna array plane, which implies that the phase front of the signal is parallel to the antenna array plane. This means that the received desired signal in the two antennas will have the same phase.

The undesired jamming signal is coming from another direction, with the angle α relative to the normal of the antenna array plane. In this case, the jamming signal will reach the antennas at different times, and there will be a phase shift between the jamming signal received by antenna A_1 compared to the signal received by A_2. Assuming l is much smaller than the distance to the signal source, it is easily shown that the difference in phase between the two will be

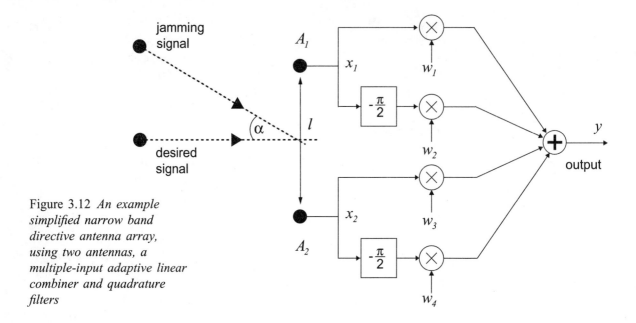

Figure 3.12 *An example simplified narrow band directive antenna array, using two antennas, a multiple-input adaptive linear combiner and quadrature filters*

$$\phi = \frac{2\pi l \sin(\alpha)}{\lambda} = \frac{\omega l \sin(\alpha)}{c} \tag{3.36}$$

where c is the propagation speed of the signal, in this case equal to the speed of light, and ω is the angular frequency. Hence, the compound signals received by antennas A_1 and A_2 respectively can be expressed as

$$x_1 = s \cos(\omega t) + u \cos(\omega t) \tag{3.37}$$

$$x_2 = s \cos(\omega t) + u \cos(\omega t - \phi) \tag{3.38}$$

The signal processing system consists of a four-input adaptive linear combiner, having the weights w_1 to w_4. The signal coming from each antenna is branched and passed directly to the combiner in one branch and via a **quadrature filter** (a Hilbert transform) in the other. The quadrature filter is an all-pass filter, introducing a $-\pi/2$ radian phase shift of the signal, but not changing the amplitude. The quadrature filter will be further discussed in Chapter 5. The output of the combiner can now be written as

$$y = w_1\big(s \cos(\omega t) + u \cos(\omega t)\big)$$

$$+ w_2 \left(s \cos\left(\omega t - \frac{\pi}{2} \right) + u \cos\left(\omega t - \frac{\pi}{2} \right) \right)$$

$$+ w_3 \left(s \cos(\omega t) + u \cos(\omega t - \phi) \right)$$

$$+ w_4 \left(s \cos\left(\omega t - \frac{\pi}{2} \right) + u \cos\left(\omega t - \phi - \frac{\pi}{2} \right) \right) \tag{3.39}$$

Using the fact that $\cos(\beta - \pi/2) = \sin(\beta)$ and rearranging the terms we obtain

$$y = s \left((w_1 + w_3) \cos(\omega t) + (w_2 + w_4)) \sin(\omega t) \right) + u \left(w_1 \cos(\omega t) \right.$$

$$\left. + w_3 \cos(\omega t - \phi) + w_2 \sin(\omega t) + w_4 \sin(\omega t - \phi) \right) \qquad (3.40)$$

Now, the objective is to cancel the jamming signal u, i.e. to create a notch in the directivity pattern of the antenna array in the direction of the jammer. Hence, we want the jamming signal term of equation (3.40) to be zero, i.e.

$$u \left(w_1 \cos(\omega t) + w_3 \cos(\omega t - \phi) + w_2 \sin(\omega t) + w_4 \sin(\omega t - \phi) \right) = 0$$

$$(3.41)$$

Using the well-known trigonometric relations: $\cos(\beta-\gamma) = \cos(\beta) \cos(\gamma) + \sin(\beta)\sin(\gamma)$ and $\sin(\beta-\gamma) = \sin(\beta) \cos(\gamma) - \cos(\beta) \sin(\gamma)$ equation (3.41) can be rewritten as

$$u \left(w_1 + w_3\cos(\phi) - w_4\sin(\phi) \right) \cos(\omega t)$$

$$+ u \left(w_2 + w_3 \sin(\phi) + w_4 \cos(\phi) \right) \sin(\omega t) = 0 \qquad (3.42)$$

In this case, solutions, i.e. the optimum weights W^*, can be found by inspection. There are many possible solutions

$$w_1 = -w_3\cos(\phi) + w_4\sin(\phi) \qquad (3.43a)$$

$$w_2 = -w_3\sin(\phi) - w_4\cos(\phi) \qquad (3.43b)$$

We choose a solution that is simple from a practical point of view, by setting

$$w_1^* = 1 \quad \text{and} \quad w_2^* = 0$$

Hence, two weights and one quadrature filter can be omitted from the structure shown in Figure 3.12. Antenna A_1 is simply connected directly to the summing point and the hardware (software) that constitutes the weights w_1 and w_2 and the upper quadrature filter can be removed.

Using equations (3.43a) and (3.43b) the components of one optimum weight vector W^* can be calculated

$$w_1^* = 1$$

$$w_2^* = 0$$

$$w_3^* = -\cos(\phi)$$

$$w_4^* = \sin(\phi)$$

Inserting W^* in equation (3.40) the output will be

$$y = s \left((1 - \cos(\phi))\cos(\omega t) + \sin(\phi)\sin(\omega t) \right)$$

$$= s(\cos(\omega t) - \cos(\omega t + \phi)) \tag{3.44}$$

The result is not surprising. The jamming signal is cancelled and the output is the sum of two components, the original signal received by antenna A_1 and a phase-shifted version of the signal received at antenna A_2. As expected, the phase shift is exactly the shift needed to cancel the jamming signal.

Now, in this case it was quite easy to find a solution analytically. In a more complex system, having for instance 30 or even more antennas and a number of interfering, jamming signals coming from different directions, there is no guarantee that an analytical solution can be found. In such a case, an adaptive system can iteratively find an optimum solution in the mean-square sense, using for instance some kind of LMS-based adaptation algorithm. Further, if the signal sources are moving, an adaptation system would be required.

Working with wide band signals, the input circuits of the linear combiner may consist not only of quadrature filters, but also more elaborate filters, e.g. adaptive FIR filters.

In this chapter only a few applications of adaptive digital processing systems have been discussed. There are numerous areas (Widrow and Stearns, 1985), for instance in process identification, modelling and control theory, where adaptive processing is successfully utilized.

4 Non-linear applications

4.1 General

There is an almost infinite number of non-linear signal processing applications. In this chapter, a few examples like **the median filter**, **artificial neural networks** (ANN) and **fuzzy logic** will be discussed. Some of these examples are devices or algorithms that are quite easy to implement using digital signal processing techniques, but would be very hard or almost impossible to build in practice using classical 'analog' methods.

4.2 The median filter

4.2.1 Basics

A median filter is a non-linear filter used for signal smoothing. It is particularly good for removing **impulsive type noise** from a signal. There are a number of variations of this filter, and a two-dimensional variant is often used in digital image processing systems to remove noise and speckles from images. The non-linear function of the median filter can be expressed as

$$y(n) = \text{med}\left[x(n-k), x(n-k+1), \cdots x(n), \cdots x(n+k-1), x(n+k)\right] \quad (4.1)$$

where $y(n)$ is the output and $x(n)$ the input signal. The filter 'collects' a window containing $N = 2k+1$ samples of the input signal and then performs the **median** operator on this set of samples. Taking the median means to sort the samples by magnitude and then select the mid-value sample (the median). For this reason, N is commonly an odd number. If for some reason an even number of samples must be used in the window, the median is defined as shown below. Assuming the samples are sorted in such a way that x_1 is the smallest and x_{2k+1} the largest value

$$\text{med}\left[x_1, x_2 \cdots x_N\right] = \begin{cases} x_{k+1} & \text{for } N = 2k+1 \\ \frac{1}{2}(x_k + x_{k+1}) & \text{for } N = 2k \end{cases} \quad (4.2)$$

Different methods have been proposed to analyse and characterize median filters. The technique of **root signal** analysis (Mitra and Kaiser, 1993) deals with signals that are invariant to median filtering and defines the 'passband' for median filters. A root signal of a median filter of length $N = 2k+1$ is a signal that passes through the filter unaltered. This signal satisfies the following equation

$$x(n) = \text{med}\left[x(n-k), x(n-k+1), \cdots x(n), \cdots x(n+k-1), x(n+k)\right] \quad (4.3)$$

The median filter itself is simple and in the standard form there is only one design parameter, namely the filter length $N = 2k+1$. There are some terms used which pertain to root signals (Gallagher Jr and Wise, 1981). A **constant neighbourhood** is a region of at least $k+1$ consecutive, identically valued points. An **edge** is a monotonically rising or falling set of points surrounded on both sides by constant neighbourhoods. An **impulse** is a set of at least one, but less than $k+1$, points whose values are different from the surrounding regions and whose surrounding regions are identically valued constant neighbourhoods. So, in essence, a **root signal** is a signal consisting of only constant neighbourhoods and edges. This definition implies that a signal which is a root signal to a median filter of length N is also a root signal of any median filter whose length is less than N. It is also interesting to note that a median filter preserves edges, both positive and negative, provided they are separated by a constant neighbourhood. The longer the filter length, the farther apart the edges have to be, but the actual magnitude of the slopes is irrelevant. This means that a median filter filters out impulses and oscillations, but preserves edges. This will be demonstrated below.

4.2.2 Threshold decomposition

Analysing combinations of linear filters by using the principle of superposition is in many cases easier than analysing combinations of non-linear devices like the median filter. However, by using a method called **threshold decomposition** we divide the analysis problem of the median filter into smaller parts.

Threshold decomposition of an integer M-valued signal $x(n)$, where $0 \leqslant x(n) < M$, means decomposing it into $M-1$ **binary** signals $x^1(n)$, $x^2(n)$, $\cdots$ $x^{M-1}(n)$

$$x^m(n) = \begin{cases} 1 & \text{if } x(n) \geqslant m \\ 0 & \text{else} \end{cases} \tag{4.4}$$

Note! The upper index is only an index, and does not imply 'raised to'. Our original M-valued signal can easily be reconstructed from the binary signal by adding them together

$$x(n) = \sum_{m=1}^{M-1} x^m(n) \tag{4.5}$$

Now, a very interesting property of a median filter (Fitch *et al.*, 1984) is that instead of filtering the original M-valued signal, we can decompose it into $M-1$ 'channels' (4.4) each containing a **binary** median filter. Then we can add the outputs of all the filters (4.5) to obtain an M-valued output signal (see Figure 4.1).

The threshold decomposition method is not only good for analysis, it is also of great interest for implementing median filters. A binary median filter is easy to implement, since the median operation can be replaced by a simple vote of majority. If there are more 'ones' than 'zeros', the filter output should be 'one'. This can be implemented in many different ways. A binary median filter of length $N = 3$, can for example be implemented using simple Boolean functions

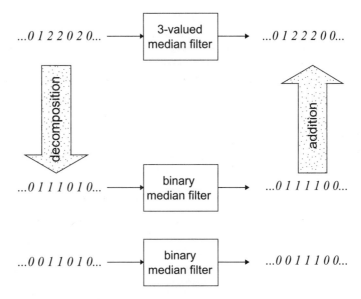

Figure 4.1 *Median filtering of a three-valued signal by threshold decomposition, M=3*

$$y(n) = x(n-1) \cap x(n) \cup x(n-1) \cap x(n+1) \cup x(n) \cap x(n+1) \quad . \qquad (4.6)$$

where $x(n)$ and $y(n)$ are binary variables and the Boolean operation $\cap$ is AND and $\cup$ is OR. For large filter lengths, this method may result in complex Boolean calculations, and a simple counter can be used as an alternative. The pseudo code below shows one implementation example.

```
binmed:    counter:=0
           for i:=1 to N do
               if x[i] then counter++
               else counter--
           if counter>0 then y[n]:=1
           else y[n]:=0
```

Now, one may argue, if there is a large number of thresholds M, there is going to be a number of binary filters as well. This is of course true, but if the filter is implemented as software for a digital signal processor, the same piece of computer code can be reused in all binary filters (as a function or sub-routine). Hence, the problem deals with the processing speed. Depending both on the kind of signals being filtered and the features of the output signal of primary interest, we may not need to process all thresholds and non-equidistant thresholds may be used.

The technique of threshold decomposition and using binary, Boolean filters has led to a new class of filters, so-called **stacked filters**.

4.2.3 Performance

If a median filter is compared to a conventional linear mean filter, implemented as an FIR filter having constant tap weights (moving average), the virtues of the median filter (running median) become apparent. Consider the following example, where we compare the output of a moving average filter of length $N = 9$ and a median filter of the same length. The mean filter with constant weights (a linear smoothing filter) can be expressed as

$$y_1(n) = \sum_{i=1}^{9} \frac{1}{9} x(n+i-5) \tag{4.7a}$$

and the median filter (a non-linear smoothing filter) as

$$y_2(n) = \text{med}\left[x(n-4), \cdots x(n), \cdots x(n+4)\right] \tag{4.7b}$$

Assume that the input signal is a square wave signal that is to be to used for synchronization purposes, i.e. the edges of the signal are of primary interest, rather than the absolute level. Unfortunately, the signal is distorted

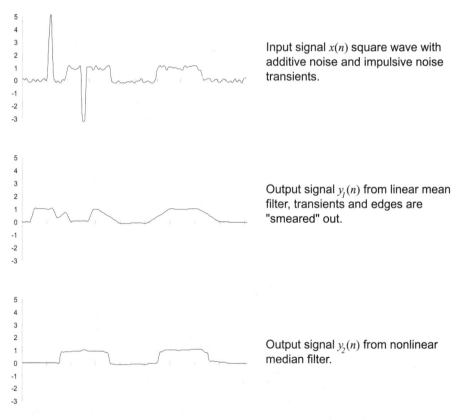

Input signal $x(n)$ square wave with additive noise and impulsive noise transients.

Output signal $y_1(n)$ from linear mean filter, transients and edges are "smeared" out.

Output signal $y_2(n)$ from nonlinear median filter.

Figure 4.2 *Input signal (top) filtered by linear running average filter (middle) and non-linear running median filter (bottom). Filter length N = 9 in both cases*

by additive Gaussian noise and impulsive type noise transients, typically generated by arcs and other abrupt discharge phenomena (see Figure 4.2).

As can be seen from Figure 4.2, when comparing the outputs of the two filters, the median filter performs quite well preserving the edges of the signal but suppressing the noise transients. The standard linear mean filter however suffers two basic problems. First, when a quite large transient occurs in the filtering window, it will affect the output as long as it is in the averaging window. This means that a narrow but strong peak, will be 'smeared' out and will basically create a new pulse that can be mistaken for a valid square wave signal. Due to the long time constant of the linear FIR filter, the edges of the desired signal will also be degraded. This may be a significant problem if timing is crucial, e.g. when the signal is used for synchronization purposes. Using a hard limiter-type device to 'sharpen up' the edges again is of course possible. However, the slope of the edges does not only depend on the response time of the mean filter, but also on the actual (varying) amplitude of the input signal. This will again result in uncertainty of the true timing.

The uncertainty of the edges in the median filter case is mainly due to the additive noise, which will be suppressed by the filter. The signal form is quite similar to the input signal with the addition of noise. For a square wave without noise, the median filter will present a 'perfect' edge.

4.2.4 Applications

If the median filter is implemented as software on a digital computer, any standard sorting algorithm like 'bubblesort', 'quicksort' etc. (Wirth, 1976) can be used to sort the values (if a stacked filter approach is not taken). When calculating the expected sorting time, it should be noted that we do not need to sort all the values in the filter window of the median filter. We only need to continue sorting until we have found the mid-value sample.

The use of median filters was first suggested for smoothing statistical data. This filter type has however, found most of its applications in the area of digital image processing. An edge preserving-filter like the median filter can remove noise and speckles without blurring the picture. Removing artifacts from imperfect data acquisition, for instance horizontal stripes sometimes produced by optical scanners, is done successfully using median filters. Median filters are also used in radiographic systems, in many commercial tomographic scan systems and for processing EEG signals and blood pressure recordings. This type of filter is also likely to be found in commercial digital TV sets in the future, because of the very good cost-to-performance ratio.

4.3 Artificial neural networks

4.3.1 Background

'**Neural networks**' is a somewhat ambiguous term for a large class of massively parallel computing models. The terminology in this area is pretty confused; scientific well-defined terms are sometimes mixed with trademarks and sales bull. A few examples are: 'connectionist's net', 'artificial neural systems (ANS)', 'parallel distributed systems (PDS)', 'dynamical functional systems', 'neuromorphic systems', 'adaptive associative networks', 'neuron computers' etc.

4.3.2 The models

In general, the models consists of a large number of typically non-linear computing **nodes**, interconnected with each other via an even larger number of adaptive **links** (weights). Using more or less crude models, the underlying idea is to mimic the function of neurons and nerves in a biological brain.

Studying neural networks may have two major purposes: either we are interested in modelling the behaviour of biological nerve systems, or we want to find smart algorithms to build computing devices for technical use. There are many interesting texts dealing with neural networks from the angle of perception, cognition and psychology (Rumelhart and McClelland, 1987; McLelland and Rumelhart, 1986; Grossberg, 1987a, b; Hinton and Anderson, 1981). In this book however, only the 'technical' use of neural network models will be treated, hence the term **'artificial neural networks'** will be used from here on.

'Computers' that are built using artificial neural network models, have many nice features:

- the 'programming' can set the weights to appropriate values. This can be done adaptively by 'training' (in a similar way to teaching humans). No procedural programming language is needed. The system will learn by examples
- the system will be able to generalize. If an earlier unknown condition occurs, the system will respond in the most sensible way based on earlier knowledge. A 'conventional' computer would simply 'hang' or exhibit some irrelevant action in the same situation
- The system can handle incomplete input data and 'soft' data. A conventional computer is mainly a number cruncher, requiring well-defined input figures
- The system is massively parallel and can easily be implemented on parallel hardware, thus obtaining high processing capacity
- The system will contain a certain amount of redundancy. If parts of the system are damaged, the system may still work, but with degraded performance ('graceful descent'). A 1 bit error in a conventional computer will in many cases cause the computer to 'crash'.

Systems of the type outlined above have been built for many different purposes and in many different sizes during the past 50 years. Some examples are systems for classification, pattern recognition, content addressable memories (CAM), adaptive control, forecasting, optimization and signal processing. Since appropriate hardware is still not available, most of the systems have been implemented in software on conventional sequential computers. This unfortunately implies that the true potential of the inherently parallel artificial neural network algorithms has not been very well exploited. The systems built so far have been pretty slow and small, typically of the size of 10^4–10^6 nodes having 10^1–10^2 links each. Taking into account that the human brain has some 10^{10}–10^{11} nodes with 10^3–10^6 links each, it is easy to realize that the artificial systems of today are indeed small. In terms of complexity, they are comparable to the brain of a fly.

4.3.3 Some historical notes

The history of artificial neural networks can be traced back to early in the twentieth century. The first formal model was published by W.S. McCulloch and W. Pitts in 1943. A simple node type was used and the network structures were very small. It was however possible to show that the network could actually perform meaningful computations. The problem was finding a good way of 'training' (i.e. 'adapting' or 'programming') the network. In 1949, D. Hebb published *The Organization of Behavior*, where he presented an early version of a correlation-based algorithm to train networks. This algorithm later came to be known as **Hebb's rule**. In the original paper, the algorithm was not well analysed or proved, so it came to be regarded as an unproved hypothesis until 1951. In the same year, M. Minsky and D. Edmonds built a 'learning machine', consisting of 300 electron tubes and miscellaneous surplus equipment from old bombers which illustrated a practical example.

It is worth noting that the artificial neural network ideas are as old as the digital computer (e.g. ENIAC (electronical numerical integrator and computer) in 1944). The digital computer was however developed faster, mainly because it was easier to implement and analyse.

In the late 1950s and early 1960s, B. Widrow and M.E. Hoff introduced a new training algorithm called the **Delta rule** or the **Widrow–Hoff algorithm**. This algorithm is related to the LMS algorithm used in adaptive digital filters. Widrow also contributed a new network node type called the **ADALINE** (adaptive linear neuron) (Widrow and Lehr, 1990). At about the same time, F. Rosenblatt worked extensively with a family of artificial neural networks called **perceptrons**. Rosenblatt was one of the first researchers to simulate artificial neural networks on conventional digital computers instead of building them using analog hardware. He formulated 'the perceptron convergence theorem' and published his *Principles of Neurodynamics* in 1962.

M. Minsky and S.A. Papert also investigated the perceptron, but they were not as enthusiastic as Rosenblatt. In 1969 they published the book *Perceptrons* (Minsky and Papert, 1969), which was a pessimistic and maybe somewhat unfair presentation, focusing on the shortcoming of the perceptrons. Unfortunately, this book had a depressing impact on the entire artificial neural network research society. Funding dried up, and for the next decade only a few researchers continued working in this area. One of them was S. Grossberg, working with 'competitive learning'. In general however, not very much happened in the area of artificial neural networks during the 1970s. Conventional digital von Neumann-based computers and symbol-oriented artificial intelligence (AI) research dominated.

At the beginning of the 1980s, a renaissance took place. The artificial neural network was 'reinvented' and a number of new and old research groups started working on the problems again, this time armed with powerful digital computers. In 1980 **'feature maps'** were presented by T. Kohonen (Finland) and in 1982 J. Hopfield published a couple of papers on his **Hopfield net**. Perhaps too well marketed, this network was able to find solutions to some famous non-polynominal-complete optimization problems. In 1983, S. Kirkpatrick *et al.* introduced **'simulated annealing'**, a way of increasing the chances to find a **global optimum** when using artificial neural networks in optimization applications.

In 1984, P. Smolensky presented 'harmony theory', a network using probabilistic node functions. In the same year, K. Fukushima demonstrated his 'neocognitron', a network which was able to identify complex handwritten characters, for example Chinese symbols.

The network training method **'back-propagation'** had been around in different versions since the late 1960s, but was further pursued by D. Rumelhart *et al.* in 1985. This year, the **'Boltzmann machine'** a probabilistic type network was also presented by G.E. Hinton and T.J. Sejnowski.

In the following years, many new variants of artificial neural network systems and new areas of application occurred. Today the area of artificial neural networks, an extremely interdisciplinary technology, is used for instance in (Lippmann, 1987; Chichocki and Unbehauen, 1993) signal processing, optimization, identification, estimation, prediction, control, robotics, data bases, medical diagnostics, biological classification (e.g. blood and genes), chemistry and economy. There are however not very many 'new' ideas presented today, rather extensions and enhancements of old theories. Unfortunately, we still lack the ideal hardware. Extensive work in the area of ASIC (application specific integrated circuit) is in progress, but there are some basic problems which are hard to overcome when trying to build larger networks. Many of these problems have to do with the large number of weights (links). For example, how should all the weights be stored and loaded into a chip? How can the settings be found, in other words, how can the training be performed? How should these examples be chosen since a large number of weights will require an even larger number of training examples?

4.3.4 Feedforward networks

The class of feedforward networks is characterized by having separate inputs and outputs and no internal feedback signal paths. Hence, there are no stability problems. The nodes of the network are often arranged in one or more discrete **layers** and are commonly easy to implement. If the nodes do not possess any dynamics like integration or differentiation, the entire network mainly performs a non-linear mapping from one vector space to another. After the training of the net is completed, meaning that the weights are determined, no iteration process is needed. A given input vector results in an output vector in a one-step process.

4.3.4.1 Nodes

The node j in a feedforward artificial neural network has a basic node function of the type

$$x_j = g\left(f\left(\sum_{i=1}^{N} w_{ij}x_i + \phi_j \right) \right) \tag{4.8}$$

where x_j is the output of node j and ϕ_j is the bias of node j. The N inputs to node j are denoted x_i where $i = 1, 2 \dots N$ and the weights (links) are w_{ij} from input i to node j. For pure feedforward networks $w_{ii} = 0$. The function f( ) is the **activation function** sometimes also called the 'squashing function'. This function can be chosen in a variety of ways, often depending on the ease of implementation. The function should however be **monotonically non-decreasing** to avoid unambiguous behaviour. Some training methods will also require the activation function to be **differentiable**. The activation function is commonly a non-linear function, since a linear function will result in a trivial network. Some examples of common activation functions are

Hard limiter

$$f(x) = \begin{cases} a & x \geq 0 \\ b & x < 0 \end{cases} \tag{4.9}$$

where a and b are constants. If $a = 1$ and $b = -1$ (4.9) turns into the sign function.

Soft limiter

$$f(x) = \begin{cases} a & x \geq a \\ x & b \leq x < a \\ b & x < b \end{cases} \tag{4.10}$$

The function is linear in the range $a \dots b$, where a and b are constants. The function saturates upwards at a and downwards at b.

Sigmoid ('logistic function')

$$f(x) = \frac{1}{1 + e^{-x/T}} \tag{4.11}$$

In this expression, the parameter T is referred to as the 'computational temperature'. For small temperatures, the function 'freezes' and the shape of equation (4.11) approaches the shape of a hard limiter equation (4.9). Sometimes the function $f(x) = \tanh(x/T)$ or similar is also used.

Finally, the **output function** g( ) may be an integration or summation or alike if the node is supposed to have some kind of memory. In most feedforward networks however, the nodes are without memory, hence a common output function is

$$g(x) = x \tag{4.12}$$

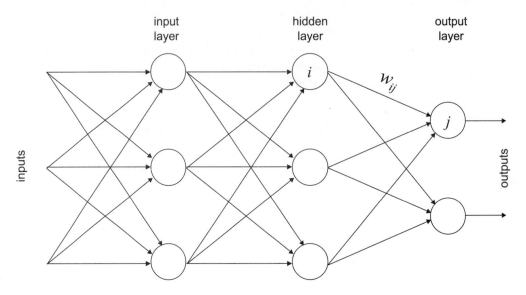

Figure 4.3 *A simple three-layer feedforward network*

4.3.4.2 Network topology

When dealing with networks consisting of a number of nodes N, matrix algebra seems to be handy in general. All weights w_{ij} in a network can for instance be collected into a weight matrix W ($N \times N$), and all signals into a vector X ($N \times 1$). The j-th column vector W_j of matrix W hence represents the weights of node j. Now, it is straightforward to see that a node basically performs a non-linear scalar product between the corresponding weight vector and the signal vector (assuming the output function is as in equation (4.12))

$$x_j - f(X^T W_j + \phi_j) \tag{4.13}$$

Using a weight matrix in this way allows arbitrary couplings between **all** nodes in the network. Unfortunately, the size of the matrix grows quickly as the number of elements increases as N^2. When dealing with feedforward type networks, this problem is counteracted by building the network in **layers** thereby reducing the number of allowed couplings. A layered network is commonly divided into three (or more) layers denoted as the input layer, the hidden layer(s) and the output layer. Figure 4.3 shows a simple three-layer feedforward network. In this structure, the input signals (input vector) enters the network at the input layer. The signals are then only allowed to proceed in one direction to the hidden layer and then finally to the output layer, where the signals (output vector) exit the network. In this way, the weight matrix can be divided into three considerably smaller ones, since the nodes in each layer only need to have access to a limited number of the signal components in the total signal vector X. Questions arise with regard to the number of layers and the number of nodes in each of these layers that should be used. While the number of nodes per layer is hard to determine, the number of layers can be calculated by the following method.

Using a layered structure, impose some restrictions on the possible mappings from the input vector space to the output vector space. Consider the following example. Assume that we have a two-dimensional input vector and require a scalar output. We use two signal levels '1' and '0'. The output should be '1' if **one** of the two input vector components is '1'. The output should be '0' otherwise. This is a basic exclusive-or function (XOR) or modulo-2 addition or 'parity' function. The task of the artificial neural network is to divide the two-dimensional input vector space into two **decision regions**. Depending in which decision region the input vector is located, a '1' or a '0' should be presented at the output.

Starting out with the simplest possible one-layer network, consisting of one node with a hard limiter-type activation function, we can express the input–output function

$$x_3 = f\left(\sum_{i=1}^{3} w_{i3}x_i + \phi_3\right) = \begin{cases} 1 & w_{13}x_1 + w_{23}x_2 + \phi_3 \geqslant 0 \\ 0 & w_{13}x_1 + w_{23}x_2 + \phi_3 < 0 \end{cases} \qquad (4.14)$$

Assume the hard limiter has the function

$$f(x) = \begin{cases} 1 & x \geqslant 0 \\ 0 & x < 0 \end{cases} \qquad (4.15)$$

The desired response is shown in the truth table shown in Table 4.1.

Now, the task is to find the weights and the bias needed to implement the desired function. It is possible to implement an AND or OR function, but an XOR function cannot be achieved. The reason is that a node of this type is only able to divide the input vector space with a (hyper)plane. In our example (from equation (4.14)) the borderline is

Table 4.1 *Truth table for XOR function*

x_1	x_2	x_3
0	0	0
0	1	1
1	0	1
1	1	0

$$x_2 = -\frac{\phi}{w_{23}} - \frac{w_{13}}{w_{23}}x_1 \qquad (4.16)$$

In other words, we cannot achieve the required shape of the decision regions, since the XOR function requires two disjunct areas like A and A (see Figure 4.4). Since many more complex decision tasks can be traced back to the XOR function, this function has a fundamental importance in computation. The basic perceptron presented by Rosenblatt was mainly a one-layer network of the type presented in the example above. The main criticism presented in the book *Perceptrons* by Minsky and Papert (Minsky and Papert, 1969), concerned the inability of the perceptron to solve the basic XOR problem. This was a bit unfair, since the perceptron can easily be modified to handle the XOR problem. This can be done by adding a second layer of nodes to our system. With respect to the example above that means adding another two nodes, and obtaining a **two-layer** artificial neural network. The network contains two nodes in the input layer, no hidden layers and one node in the output layer. The nodes are of the same simple type as in the example above. Now, each of the two nodes in the input layer can divide the input vector space with hyperplanes. In our two-dimensional example, this is two straight lines as in equation (4.16). If one node has its decision line to the left of the two desired areas A (see Figure 4.4), the

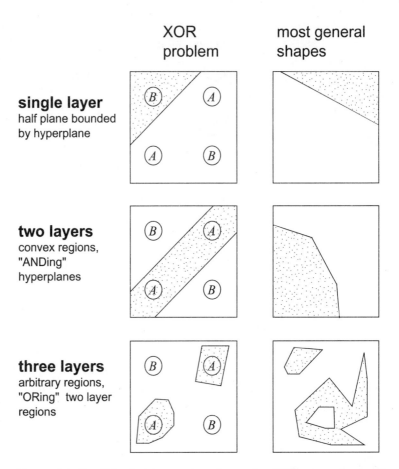

Figure 4.4 *Possible decision region shapes for different numbers of layers*

output of the node will be '1' when an input vector corresponding to A is present. The other node has its line to the right, of the desired decision region and gives a '1' output signal for A. The desired response can now be obtained by 'ANDing' the output of the two nodes. AND and OR functions can easily be achieved using a single node, so the node in the output layer of our example network will do the 'ANDing' of the outputs from the nodes in the input layer. Hence, the XOR problem is solved.

Figure 4.5 shows a proposed solution to the XOR problem. Since the input signals have the indices 1 and 2 respectively, the nodes are numbered as 3 and 4 in the input layer and 5 in the output layer. The weights and biases of the nodes may be chosen as

$$\text{node 3:}\quad w_{13} = 1 \quad w_{23} = 1 \quad \phi_3 = -0.5$$
$$\text{node 4:}\quad w_{14} = -1 \quad w_{24} = -1 \quad \phi_4 = 1.5$$
$$\text{node 5:}\quad w_{45} = 1 \quad w_{35} = 1 \quad \phi_5 = -1.5$$

Note! There are many possible solutions, try it yourself.

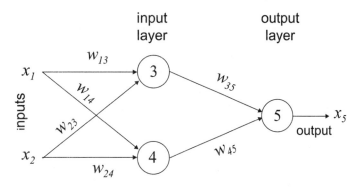

Figure 4.5 *Example of a two-layer feedforward artificial neural network of perceptron type, capable of solving the XOR problem*

Going back to Figure 4.4, we can now draw some general conclusions about the possible shapes of decision regions as a function of the number of layers. A one-layered network can only split the input vector space with hyperplanes. Using a second layer, having the outputs of the first layer as inputs, a number of hyperplanes can be 'ANDed' together and convex decision regions can now be formed. The number of nodes in each layer determines the number and complexity of the regions. Finally, incorporating a third layer, a number of convex regions, created by the preceding second layer, can now be 'ORed' together and arbitrary complex decision regions can be obtained. So, from a theoretical point of view, there is no need for more than three layers in a layered feedforward artificial neural network. When implementing these nets however, more layers may sometimes be used. This may facilitate reuse of some intermediate computing results and reduce the total number of nodes in the network and/or simplify the training, that is, the process of determining the weights and biases in the nodes.

Returning to the XOR example, there is another way of solving the problem using only **one** perceptron-type node. The underlying principle is the way in which the input signals or 'features' are chosen. This is a very important, often forgotten topic in many cases. Choosing 'bad' input signals may make it impossible even for an advanced artificial neural network to perform a simple task, while choosing 'smart' input signals complex problems makes it possible to solve using fairly simple networks. If we put a non-linear 'preprocessor' in front of the inputs of our original single layer network, creating for instance the two input signals

$$y_1 = x_1 + x_2$$
$$y_2 = (x_1 + x_2)^2 \tag{4.17}$$

the XOR problem can now be solved using the original single perceptron node (index 3) with the weights and bias

node 3: $w_{13} = 2$ $w_{23} = -1$ $\phi_3 = -0.5$

Note that y_2 represents a circular area, i.e. a convex region.

The importance of a wise selection of input signals or features for a given problem and network cannot be overemphasized.

4.3.4.3 Training and adaptation

Training or adaptation is to determine the weights and biases of an artificial neural network, i.e. to determine the function of the network. By means of a number of algorithms this can be done in a number of ways. Before going into a more detailed discussion, some general problems should be addressed.

Training by examples is of course a nice way of 'programming' the network. However, the more weights and biases that need to be determined, the more input data examples will be needed (and the more training time will be required). Adaptive systems like this 'eat' input data. Once used as an example, the data is normally not good for any further training.

Secondly, finding a significant sub-set of all possible input data suited for training is not easy in the general case. Using a 'bad' training set, the network may 'learn' the wrong mappings and poor performance may result (like teenagers?). Normally, training is performed using a **training set** of input data, while performance testing uses another set of input data. If the same input data set is used for both testing and training, we can only be sure that the network has learnt how to handle the **known** data set properly. If this is the primary task of the system, it would probably be better to use a lookup table algorithm (LUT) than an artificial neural network.

One of the main advantages of an artificial neural network is that it is able to generalize, i.e. if earlier **unknown** input data is presented, the network can still give a 'reasonable' answer. A 'conventional' computing system would, in many cases, present a useless error message in the same situation. The network can in this respect be viewed as a huge lookup table, where only a fraction of all the possible entries have been initialized (by the training data set). The network itself will then 'interpolate' all the non-initialized table entries. Hence, a proper choice of training data is crucial to achieve the desired function.

Training methods can be divided into two groups, **unsupervised** and **supervised.** In both cases, the network being trained is presented with training data. In the unsupervised (Rumelhart and McClelland, 1987) case, the network itself is allowed to determine a proper output. The rule for this is built into the network. In many cases, this makes the network behaviour hard to predict. This type of network is known as a 'clustering' device. The problem with unsupervised learning is that the result of the training is somewhat uncertain. Sometimes the network may come up with training results that are almost incomprehensible. Another problem is that the order in which the training data is presented may affect the final training result.

In the case of supervised training, not only is the training data presented to the network, but also the corresponding desired output. There are many different supervised training algorithms around, but most of them stem from a few classical ones, which will be presented below. The bias term ϕ of a node can be treated as a weight connected to an input signal with a constant value of 1.

The oldest training algorithm is probably **Hebb's rule**. This training algorithm can be used for single nodes. There are many extensions to this rule of how to update the weights, but in its basic form it can be expressed as the following

$$w_{ij}(n+1) = w_{ij}(n) + \mu x_i(n)x_j(n) \qquad (4.18)$$

where μ is the 'learning rate', x_i is the output of node i, which is also one of the inputs to node j and x_j is the output of node j. The underlying idea of Hebb's rule is that if both node i and node j are 'active' (positive output) or 'inactive' (negative output) the weight in between them, w_{ij}, should be increased, else it should be decreased. In a one-layer network, x_i is a component in the supervised training input data vector and x_j corresponds to the desired output.

This simple rule has two advantages. It is **local**, that is it only needs local information x_i and x_j, and it does not require the non-linearity f( ) of the node to be differentiable. A drawback is that to obtain good training results, the input vectors in the training set need to be **orthogonal**. This is because all the training input vectors are added to the weight vectors of the active nodes, hence two or more nodes may obtain almost the same weight vectors if they become 'active' often. Further, the magnitude of the weight vectors may grow very large. This will result in 'cross talk' and erroneous outputs. If on the other hand the input data vectors are orthogonal, it is very unlikely that two nodes will be active about equally often and hence obtain similar weight vectors.

The **Widrow–Hoff rule** or **delta rule** is closely related to the LMS algorithm discussed in Chapter 3. This training algorithm can be used for single nodes in single-layer networks.

In this algorithm, an error signal ε_j is calculated as (for node j)

$$\varepsilon_j = d_j - u_j = d_j - \sum_i w_{ij}x_i - \phi_j \qquad (4.19)$$

Note that u_j is what comes out of the summation stage in the node and goes into the activation function f( ). The term d_j is the desired output signal when the training data vector consisting of $x_1, x_2, \ldots x_N$ is applied. The weights are then updated using

$$w_{ij}(n+1) = w_{ij}(n) + \mu x_i(n)\varepsilon_j(n) \qquad (4.20)$$

In this case the input vectors in the training set do not need to be orthogonal. The **difference** between the actual output and the desired output, that is the error signal ε_j, is used to control the addition of the input data vector to the weight vector. This rule is also local and it does not require the non-linearity f( ) of the node to be differentiable.

The **perceptron learning rule** is a variation of the delta rule described above. This training algorithm can, in its standard form, only be used for single-layer networks. The idea is to adjust the weights in the network in such a way as to minimize the total sum of square errors. The total error for the network is

$$E = \sum_j \varepsilon_j^2 = \sum_j (d_j - x_j)^2 = \sum_j \left(d_j - f\left(\sum_i w_{ij}x_i - \phi_j \right) \right)^2 \qquad (4.21)$$

A simple steepest descent approach is now to take the derivative of the error with respect to the weight w_{ij}. From this we can tell in what direction to change this particular weight to reduce the total square error for the network. This procedure is repeated for all weights, hence we calculate the gradient of the weight vector on the quadratic error hypersurface. For w_{ij} we obtain

$$\frac{\partial E}{\partial w_{ij}} = -2(d_j - x_j)\frac{\partial x_j}{\partial w_{ij}} = -2\varepsilon_j \frac{\partial x_j}{\partial u_j}\frac{\partial u_j}{\partial w_{ij}} = -2\varepsilon_j\, f'(u_j)x_i \qquad (4.22)$$

As can be seen, this algorithm requires the activation function f( ) of the node to be differentiable. This is the delta rule, but we have now also included the activation function. The weights are now updated as

$$w_{ij}(n+1) = w_{ij}(n) + \mu x_i(n)\varepsilon_j(n)\,f'(u_j(n)) \qquad (4.23)$$

where (as before)

$$u_j = \sum_i w_{ij}x_i + \phi_j \qquad (4.24)$$

So far, we have discussed training algorithms for single-layer feedforward artificial neural networks. Supervised training of multi-layer networks is harder, since in normal cases we lack desired output signals for all nodes that are not in the output layer.

For training multi-layer feedforward networks, there is an algorithm called **the generalized perceptron learning rule**, sometimes denoted **the back-propagation algorithm**. This algorithm is a generalization of the perceptron learning rule discussed above.

The idea is to start training the output layer. Since we do have access to desired output values d_k, this can be done by using the standard perceptron learning rule as outlined above. Now, we go backwards in the layer and start training the first layer preceding the output layer. Here we have to calculate the errors on the outputs of the hidden nodes, by propagating the errors on the output, through the output layer. This is where 'back-propagation' becomes useful.

Assume that there are M nodes in the output layer, and that we are to adjust the weights of node j in the layer just below the output layer. Firstly, the error on the output of node j will contribute to the errors of all output nodes. Let us calculate this coupling first, by taking the derivative of the total output squared error E with respect to the output of node j

$$\frac{\partial E}{\partial x_j} = \sum_{k \in M} \frac{\partial E}{\partial u_k}\frac{\partial u_k}{\partial x_j} = -2\sum_{k \in M}(d_k - x_k)\frac{\partial x_k}{\partial u_k}\frac{\partial u_k}{\partial x_j}$$

$$= -2\sum_{k \in M}(d_k - x_k)\,f'(u_k)\,w_{jk} \qquad (4.25)$$

When this link between the error on the output and the error on the output of node j is known, we can find the derivative of the total error with respect to the weights we are to update in node j (see also equation (4.22) above)

$$\frac{\partial E}{\partial w_{ij}} = \frac{\partial E}{\partial x_j}\frac{\partial x_j}{\partial w_{ij}} = \frac{\partial E}{\partial x_j}\frac{\partial x_j}{\partial u_j}\frac{\partial u_j}{\partial w_{ij}} = \frac{\partial E}{\partial x_j}f'(u_j)x_i$$

$$= -2f'(u_j)x_i\sum_{k \in M}\varepsilon_k f'(u_k)w_{jk} \qquad (4.26)$$

The weights can now be updated using the standard form (see also equation (4.23) above)

$$w_{ij}(n+1) = w_{ij}(n) + \mu x_i(n)f'(u_j(n))\sum_{k \in M}\varepsilon_k(n)\,f'(u_k(n))w_{jk} \qquad (4.27)$$

This procedure is repeated for the next layer and so on, until the weights of the nodes in the input layer have been updated.

4.3.4.4 Applications

Feedforward artificial neural networks are typically used in applications like **pattern recognition**, **pattern restoration** and **classification**. The 'patterns' are the input vectors, where the components are 'features' relevant in the application. If we are, for instance, dealing with a speech recognition system, the features can for example represent signal power in different frequency bands of a speech signal. The features can also be pixel values in an artificial neural network-based image processing system for OCR. Other areas are processing of radar and sonar echo signals, matching fingerprints, correcting transmission errors in digital communication systems, classifying blood samples, genes and ECG signals in biological and medical applications, troubleshooting of electronic systems etc. In most cases, the applications belong to some basic system types as detailed below.

Pattern associator. In this system type, the network mainly performs a mapping from the input vector space to the output vector space. The mechanism here is to map (or 'translate') an input vector to an output vector in a way that the network has been trained. If the desired mapping is simple and can be expressed algebraically, no artificial neural network is normally needed. If on the other hand the mapping cannot be easily formulated and/or only samples of input and output vectors are available, training an artificial neural network is possible.

Networks of this type have for instance been tried for weather forecasting (Hu, 1963). Since the coupling between temperature, humidity, air pressure, wind speed and direction in different places affects the weather in a very complicated way, it is easier to train a network than to formulate mathematical relations.

Auto-associator or content addressable memory (CAM). This system type can be viewed as a special case of the pattern associator above, in the sense that the input vector is mapped back on itself. The system acts as a content addressable memory, where the previously trained patterns are stored in the weights. If an incomplete or distorted version of an earlier known pattern is presented to the network, it will respond with a restored, error-free pattern.

The human memory works as a content addressable memory. Once we get some 'clues', or parts of the requested set of information, the rest of this information will be recalled by associations. Networks of this type have been tried for database searches and for error correction of distorted digital signals in telecommunication systems.

Classifier, identifiers. This system type can also be regarded as a special case of the pattern associator above. In this situation, our aim is to categorize the input pattern and to tell to what class of patterns it belongs. This type of network typically has few outputs and in some cases also a special type of output layer having 'lateral feedback' (treated below), that assures that only one output (class) at a time can be active.

Examples of this type of network are systems for classifying ECG signals (Specht, 1964), and sonar and radar echo signals. The outputs may be 'healthy/unhealthy', 'submarine/no submarine' and so on.

Regularity detectors. This system type can be viewed as a variant of the classifier above, but in this case there is no *a priori* set of categories into which the input patterns are to be classified. The network is trained using unsupervised training, that is, the system must develop its own featured representation of the input patterns. In this way, the system is used to explore statistically salient features of the population of input patterns. 'Competitive learning' (Rumelhart and McClelland, 1987) is an algorithm well suited for regularity detectors. Examples of applications in this area are finding significant parameters in a large set of data or finding good data compression and error-correcting coding schemes.

4.3.5 Feedback networks

Feedback networks, also known as **recurrent networks**, have all outputs internally fed back to the inputs in the general case. Hence, a network of this type commonly does not have dedicated inputs and outputs. Besides the non-linear activation function, nodes in such a network also have some kind of dynamics built in, for instance an integrator or accumulator.

The main idea of this class of networks is **iteration**. The input vector to this system is applied as initial states of the node outputs and/or as bias values. After these initial conditions are set, the network is iterated until convergence, when the components of the output vector can be found on the node outputs. To achieve convergence and to avoid stability problems, the weights, that is the feedback parameters, have to be chosen carefully. It is very common that the weights are constant set *a priori* in feedback networks, hence this type of network is not 'trained' in the same way as feedforward type networks.

Common networks of this type are the **Hopfield net**, the **Boltzmann machine** and **Kohonen's feature maps**.

4.3.5.1 Nodes

The node functions in a feedback network have many similarities to the node functions used in feedforward networks. An important difference however is that in the feedback network case, the node functions include dynamics or memory, for example an integrator. This is necessary as the network will be iterated to obtain the final output. (In the feedforward case, obtaining the output is a one-step process.) A common node model, used by for example Hopfield has the differential equation

$$\frac{du_j}{dt} = \sum_i w_{ij} x_i + \phi_j - u_j = \sum_i w_{ij} \mathrm{f}(u_i) + \phi_j - u_j \qquad (4.28)$$

where f( ) is the activation function. As can be seen, we have taken the node function (4.8) and inserted an integrator with negative feedback in between the summation point and the non-linear activation function. Hence, the feedback network is but a system of N non-linear differential equations, which can be expressed in a compact matrix form

$$\frac{dU}{dt} = W^T X + \Theta - U = W^T F(U) + \Theta - U \qquad (4.29)$$

where

$$U = \begin{bmatrix} u_1 & u_2 & \cdots & u_N \end{bmatrix}^T$$

$$F(U) = \begin{bmatrix} \mathrm{f}(u_1) & \mathrm{f}(u_2) & \cdots & \mathrm{f}(u_N) \end{bmatrix}^T = X$$

$$\Theta = \begin{bmatrix} \phi_1 & \phi_2 & \cdots & \phi_N \end{bmatrix}^T$$

Now, it is straightforward to realize that when iterating the network it will converge to a stable state when equation (4.29) is equal to the zero vector

$$\frac{dU}{dt} = 0 \qquad (4.30)$$

If we assume that the weight matrix W is symmetric, that is, $w_{ij} = w_{ji}$ and that we integrate equation (4.29) we can define the computational 'energy' function for the network

$$\mathrm{H}(X) = -\frac{1}{2} X^T W^T X - X^T \Theta + \sum_{j=1}^{N} \int_0^{x_j} \mathrm{f}^{-1}(x)\, dx \qquad (4.30)$$

Now, if the activation function is a sigmoid (4.11), the inverse function will be

$$u = \mathrm{f}^{-1}(x) = -T \ln \left(\frac{1}{x} - 1 \right) \qquad (4.31)$$

If we, like Hopfield, use the hard limiter activation function, this corresponds to a sigmoid with a low computational temperature T, implying that the third term of equation (4.30) will be small. Hence, it can be neglected and equation (4.30) can be simplified to

$$H(X) = -\frac{1}{2}X^T W^T X - X^T \Theta \tag{4.32}$$

Hence, when the network settles to a stable state (4.30) all derivatives, that is the gradient of the energy function (4.32), are zero. This means that the stable states of the network and the output patterns correspond to a local minimum of the energy function. To 'program' the network means formulating the problem to be solved by the network in terms of an energy function of the form function (4.32). When this energy function has been defined, it is straightforward to identify the weights and biases of the nodes in the feedback artificial neural network.

A more general way to design a feedback network is of course to start out defining an energy function and to obtain the node functions by taking the derivatives of the energy function. This method is in the general case much harder, since node functions may turn out to be complicated and convergence problems may occur. The energy function must be a Lyapunov (Åström and Wittenmark, 1984) type function, that is it should be monotonically decreasing in time. Further, Hopfield has shown that the weight matrix W must be symmetric, in other words $w_{ij} = w_{ji}$ and have vanishing diagonal elements $w_{ii} = 0$ to guarantee stability.

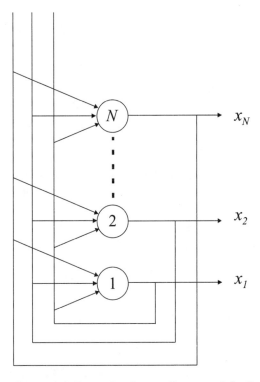

Figure 4.6 *Example of a small, general feedback artificial neural network*

4.3.5.2 Network topology

In the general case, all nodes are connected with all other nodes in a feed-back network (see Figure 4.6). There are however some special cases; one such case is **lateral feedback** (Lippmann, 1987; Stranneby, 1990). This type of feedback is commonly used within a layer of a feedforward network. An example is the **MAXNET** (maximum likelihood Hamming net). This lateral feedback type network assures that one and only one output of a layer is active and that the active node is the one having the largest magnitude of the input signal of all nodes in the layer. The MAXNET is hence a device to find the maximum value within all elements in a vector. Quite often a structure like this is used on the output of a classifying feedforward type artificial neural network, to guarantee that only one class is selected.

4.3.5.3 Local and global minimum

As was seen in the previous section, when the feedback artificial neural network is iterated, it will finally settle in one of the minimum points of the energy function of the type such as function (4.32); which minimum point depends on the initial state of the network, that is the input vector. The network will simply 'fall down' in the 'closest' minimum after iteration is started. In some applications this is a desired property, in others it is not.

If we would like the network to find the **global** minimum, in other words the 'best' solution in some sense, we need to include some extra mechanism to assure that the iteration is not trapped in the closest **local** minimum. One network having such a mechanism is the **Boltzmann machine**.

In this type of feedback artificial neural network, **stochastic node functions** are used. Every node is fed additive zero-mean independent (uncorrelated) noise. This noise has the effect of making the node outputs noisy, thus 'boiling' the hypersurface of the energy function. In this way, if the state of the network happens to 'fall down' in a local minimum, there is a certain probability that it will 'jump up' again and fall down in another minimum. The trick is then gradually to decrease the effect of the noise, to make the network finally end in the 'deepest' minimum, that is the global one. This can be performed by slowly decreasing the 'temperature' T in the sigmoid activation function in the network nodes.

The trick of decreasing the 'temperature' or cooling the network until it 'freezes' to a solution is called **simulated annealing**. The challenge here is to find the smartest annealing scheme that will bring the network to the global minimum with a high probability in as short a time as possible. Quite often networks using simulated annealing converge very slowly.

The term 'annealing' is borrowed from crystallography. At high temper-atures, the atoms of a metal lose the solid-state phase and the particles position themselves randomly according to statistical mechanics. The parti-cles of the molten metal tend towards the minimum energy state, but the high thermal energy prevents this. The minimum energy state means a highly ordered state such as a defect-free crystal lattice. To achieve defect-free crys-tals, the metal is annealed, that is, it is first heated to a temperature above the melting point and then slowly cooled. The slow cooling is necessary to prevent dislocations and other crystal lattice disruptions.

4.3.5.4 Applications

Artificial neural networks have traditionally been implemented in two ways, either by using analog electronic circuits or as software on traditional digital computers. In the latter case, which is relevant for DSP applications, we of course have to use the discrete time equivalents of the continuous time expressions present in this chapter. In the case of feedforward artificial neural networks, the conversion to discrete time is trivial. For feedback networks having node dynamics, standard numerical methods like Runge–Kutta may be used.

Feedback artificial neural networks are used in content addressable memories (CAM) and for solving miscellaneous optimization problems.

In **content addressable memories**, each minimum in the energy function corresponds to a memorized pattern. Hence, in this case, 'falling' in the closest local minimum is a desirable property. Given an incomplete version of a known pattern, the network can accomplish **pattern completion**. Unfortunately, the packing density of patterns in a given network is not very impressive. Research is in progress, in an attempt to find better energy functions that are able to harbour more and 'narrow' minimums but still result in stable networks.

Solving **optimization problems** is probably the main application of feedback artificial neural networks. In this case, the energy function is derived from the objective function of the underlying optimization problem. The global minimum is of primary interest, hence simulated annealing and similar procedures are often used. For some hard optimization problems and/or problems with real-time requirements (e.g. in control systems) a local minimum, i.e. sub-optimum solution, found in a reasonable time may be satisfactory.

Many classical NP-complete optimization problems, e.g. 'the travelling salesman' (Hopfield and Tanks, 1986) and 'the 8-queens problem' (Holmes, 1989) have been solved using feedback artificial neural networks.

It is also possible to solve optimization problems using a feedforward artificial neural network with an external feedback path. This method has been proposed for transmitter power and frequency assignment in radio networks (Stranneby, 1996).

4.4 Fuzzy logic

4.4.1 General

A **fuzzy logic** or **fuzzy control** (Passino and Yurkovich, 1998; Palm *et al.*, 1996) system performs a static, non-linear mapping between input and output signals. It can in some respects be viewed as a special class of feedforward artificial neural network. The idea was originally proposed in 1965 by Professor Lofti Zadeh at the University of California, but has not been used very much until the past decade. In ordinary logic, only false or true is considered, but in a fuzzy system we can also deal with intermediate levels, for example, one statement can be 43% true and another one 89% false.

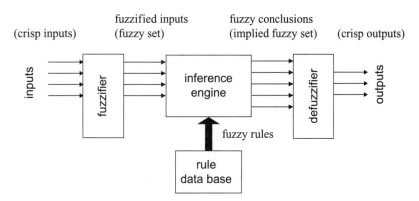

Figure 4.7 *Example of a simple fuzzy logic system*

Another interesting property is that the behaviour of a fuzzy system is not described using algorithms and formulas, but rather as a set of **rules**, that may be expressed in natural language. Hence, this kind of system is well suited in situations where no mathematical models can be formulated or when only heuristics are available. Using this approach, practical experience can be converted into a systematic, mathematical form, in for instance a control system.

A simple fuzzy logic system is shown in Figure 4.7. The **fuzzifier** uses **membership functions** to convert the input signals to a form that the **inference engine** can handle. The inference engine works as an 'expert', interpreting the input data and making decisions based on the rules stored in the **rule data base**. The rule data base can be viewed as a set of 'if–then' rules. These rules can be linguistic descriptions, formulated in a way similar to the knowledge of a human expert. Finally, the **defuzzifier** converts the conclusions made by the inference engine into output signals.

As an example, assume we are to build a smart radar-assisted cruise control for a car. The input signals to our simplified system are the *speed* of our car and the *distance* to the next car in front of us (the distance is measured using radar equipment mounted on the front bumper). The continuous output control signals are *accelerate* and *brake*. The task of the fuzzy controller is to keep the car at a constant speed, but to avoid crashing into the next car in a traffic jam. Further, to get a smooth ride, the accelerator and brake should be operated gently.

4.4.2 Membership functions

The outputs of the fuzzifier are called **linguistic variables**. In the above example we could introduce two such variables, SPEED and DISTANCE. The **linguistic value** of such a variable is described using adjectives like slow, fast, very fast etc. The task of a membership function is to interpret the value of the continuous input signal to **degree of membership** with respect to a given linguistic value. The degree of membership can be any

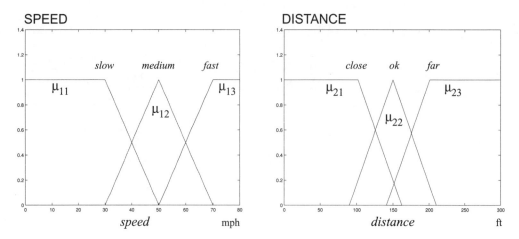

Figure 4.8 *Input signal membership functions used in the example*

number between zero and one and it can be regarded as a measure of to what extent the input signal has the property represented by the linguistic value. It is common to use 3, 5 or 7 linguistic values for every linguistic variable, hence the same number of membership functions are required. Figure 4.8 shows the membership functions for our example given above. Here we have used three linguistic values for SPEED, *slow, medium* and *fast* and three values for DISTANCE, *close, ok* and *far*.

From the figure it can be seen that the set of membership functions converts the value of the continuous input signal to degree of membership for every linguistic value. In this example, the membership functions overlap, so the SPEED can for instance be both *medium* and *fast* but with a different degree of membership. Further, we have chosen to 'saturate' the outermost membership functions. For instance, if the distance goes to plus or minus infinity it is regarded as *far* and *close* respectively. There are many possible shapes for a membership function; the simplest and most common one is a symmetric triangle, which we have used in our example. Other common shapes are trapezoid, Gaussian or similar 'bell-shaped' functions, peak shapes and skewed triangles. If a purely rectangular shape is used, the degree of membership can only be one or zero and nothing in between. In such a case, we are said to have a **crisp** set representation rather than a fuzzy set representation.

4.4.3 Fuzzy rules and inference

The input to the inference engine is the linguistic variables produced by the membership functions in the fuzzifier (the fuzzy set). The output linguistic variables of the inference engine are called the **conclusions** (the implied fuzzy set). The mapping between the input and output variables is specified in natural language by rules having the form

If premise **then** consequent (4.33)

The rules can be formulated in many forms. Of these forms, two are commonly standard: multi-input single-output (MISO) and multi-input multi-output (MIMO). In this text, we will deal with MISO rules only (a MIMO rule is equivalent to a number of MISO rules). The premise of a rule is in the multi-input case a 'logic' combination of conditions. Two common 'logic' operators are **AND** and **OR**. Two simple examples of rules containing three conditions are

If condition$_1$ **AND** condition$_2$ **AND** condition$_3$ **then** consequent (4.34a)

If condition$_1$ **OR** condition$_2$ **OR** condition$_3$ **then** consequent (4.34b)

Now, if we had been using Boolean variables (one or zero) the definition of the operators AND and OR would be obvious. In this case however, the conditions are represented by degree of membership, that is any real number between zero and one. Hence, we need an extended definition of the operators. There are different ways to define AND and OR in fuzzy logic systems. The most common way to define the AND operation between two membership values μ_{ij} and μ_{kl} is using the minimum

$$\mu_{ij} \text{ AND } \mu_{kl} = \min \{\mu_{ij}, \mu_{kl}\}$$ (4.35)

where μ_{ij} means the membership value of function j of the linguistic variable i. For example, 0.5 AND 0.7 = 0.5. The most common way to define the OR operation is using the maximum

$$\mu_{ij} \text{ OR } \mu_{kl} = \max \{\mu_{ij}, \mu_{kl}\}$$ (4.36)

For example, 0.5 OR 0.7 = 0.7. An alternative way of defining the operators is to use algebraic methods

$$\mu_{ij} \text{ AND } \mu_{kl} = \mu_{ij}\mu_{kl}$$ (4.37)

$$\mu_{ij} \text{ OR } \mu_{kl} = \mu_{ij} + \mu_{kl} - \mu_{ij}\mu_{kl}$$ (4.38)

An interesting question is how many rules are needed in the rule database? Assuming we have n linguistic variables (inputs to the inference engine) and that the number of membership functions for variable i is N_i, then the total number of rules will be

$$N_R = \prod_{i=1}^{n} N_i = N_1 \cdot N_2 \cdot \ldots \cdot N_n$$ (4.39)

This assumes we need to consider **all** possible combinations of input signals. Further, if MISO rules are used, N_R rules may be needed for every output linguistic variable in the worst case. From equation (4.39) it is easy to see that the rule database and the computational burden grow quickly if too many variables and membership functions are used. Hence, the selection of good input signals and the appropriate number of linguistic values is crucial to system performance.

To pursue our example, we need to define output linguistic variables, values and membership functions for the conclusions produced by the inference engine. Let us introduce two output variables. Firstly, ACCELERATE having three values, *release, maintain* and *press* (for the accelerator pedal). Secondly, BRAKE with the values *release, press* and *press hard* (for the brake pedal). We are assuming automatic transmission and are ignoring 'kickdown' features and so on. The associated membership functions used are triangular (see Figure 4.9). In this case, 'saturating' functions cannot be used since infinite motor power and accelerating brakes are not physically possible. From Figure 4.9 we can also see that the centre of a membership function corresponds to a given value of thrust or braking force.

ACCELERATE:	*release*	− 25% thrust	(braking using motor)
	maintain	+ 25% thrust	(to counteract air drag)
	press	+ 75% thrust	(to accelerate)
BRAKE:	*release*	0% braking force	
	press	25% braking force	(gentle stopping)
	press hard	75% braking force	(panic!)

The next step is to formulate the rules. According to equation (4.39) we will need 3·3 = 9 rules for every output (using MISO rules) if all combinations are considered. The following rules are suggested.

If SPEED is *slow* **AND** DISTANCE is *close* **then** ACCELERATE is *release* (4.40a)

If SPEED is *slow* **AND** DISTANCE is *ok* **then** ACCELERATE is *press* (4.40b)

If SPEED is *slow* **AND** DISTANCE is *far* **then** ACCELERATE is *press* (4.40c)

If SPEED is *medium* **AND** DISTANCE is *close* **then** ACCELERATE is *release* (4.40d)

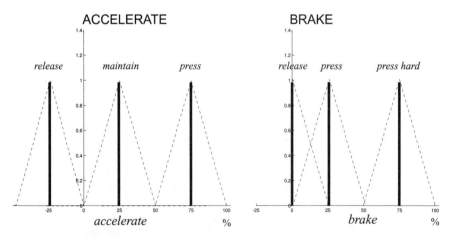

Figure 4.9 *Output signals and associated membership functions used in the example*

If SPEED is *medium* **AND** DISTANCE is *ok* **then** ACCELERATE is *maintain*

(4.40e)

If SPEED is *medium* **AND** DISTANCE is *far* **then** ACCELERATE is *maintain*

(4.40f)

If SPEED is *fast* **AND** DISTANCE is *close* **then** ACCELERATE is *release* (4.40g)

If SPEED is *fast* **AND** DISTANCE is *ok* **then** ACCELERATE is *release* (4.40h)

If SPEED is *fast* **AND** DISTANCE is *far* **then** ACCELERATE is *release* (4.40i)

If SPEED is *slow* **AND** DISTANCE is *close* **then** BRAKE is *press* (4.41a)

If SPEED is *slow* **AND** DISTANCE is *ok* **then** BRAKE is *release* (4.41b)

If SPEED is *slow* **AND** DISTANCE is *far* **then** BRAKE is *release* (4.41c)

If SPEED is *medium* **AND** DISTANCE is *close* **then** BRAKE is *press* (4.41d)

If SPEED is *medium* **AND** DISTANCE is *ok* **then** BRAKE is *release* (4.41e)

If SPEED is *medium* **AND** DISTANCE is *far* **then** BRAKE is *release* (4.41f)

If SPEED is *fast* **AND** DISTANCE is *close* **then** BRAKE is *press hard* (4.41g)

If SPEED is *fast* **AND** DISTANCE is *ok* **then** BRAKE is *press* (4.41h)

If SPEED is *fast* **AND** DISTANCE is *far* **then** BRAKE is *release* (4.41i)

Having defined all rules needed, we may now be able to find simplifications to reduce the number of rules. For instance, it is possible to reduce the rules (4.40g)–(4.40i) to just one rule, since the value of DISTANCE does not matter, in all cases ACCELERATE should be *release* if SPEED is *fast*. The reduced rule is

If SPEED is *fast* **then** ACCELERATE is *release* (4.42a)

Reasoning in the same way, rules (4.40a) and (4.40d) can be reduced to

If DISTANCE is *close* **then** ACCELERATE is *release* (4.42b)

Combining rules (4.42a) and (4.42b) we obtain

If SPEED is *fast* **OR** DISTANCE is *close* **then** ACCELERATE is *release*

(4.42c)

Keeping rules (4.40b), (4.40c), (4.40e) and (4.40f) as is, we are now left with five rules for ACCELERATE. Trying some reduction for the rules generating BRAKE, one solution is to join rules (4.41c), (4.41f) and (4.41i) to obtain

If DISTANCE is *far* **then** BRAKE is *release* (4.43a)

Leaving the other rules for BRAKE as is, we now have seven rules, so the total database will contain 12 rules (MISO). Further reductions are possible for both ACCELERATE and BRAKE, but they will result in complex rules and will probably not reduce the computational demands. Finally, the rules are rewritten in 'mathematical' form and stored in the rule database. If we adopt the following notations

μ_{11} membership value for SPEED, *slow*

μ_{12} membership value for SPEED, *medium*

μ_{13} membership value for SPEED, *fast*

μ_{21} membership value for DISTANCE, *close*

μ_{22} membership value for DISTANCE, *ok*

μ_{23} membership value for DISTANCE, *far*

for conclusion (implied fuzzy set) about thrust

η_{3i} implied membership value for ACCELERATE, by rule i

b_{3i} position of peak of recommended membership function in ACCELERATE, by rule i

for conclusion (implied fuzzy set) about braking force

η_{4j} implied membership value for BRAKE, by rule j

b_{4j} position of peak of recommended membership function in BRAKE, by rule j

we can now formulate our reduced set of rules above in a mathematical form. For ACCELERATE we get

(4.42c): $\eta_{31} = \max \{\mu_{13}, \mu_{21}\}$ for $b_{31} = -0.25$ (4.44a)

(4.40b): $\eta_{32} = \min \{u_{11}, \mu_{22}\}$ for $b_{32} = 0.75$ (4.44b)

(4.40c): $\eta_{33} = \min \{u_{11}, \mu_{23}\}$ for $b_{33} = 0.75$ (4.44c)

(4.40e): $\eta_{34} = \min \{u_{12}, \mu_{22}\}$ for $b_{34} = 0.25$ (4.44d)

(4.40f): $\eta_{35} = \min \{\mu_{12}, \mu_{23}\}$ for $b_{35} = 0.25$ (4.44e)

and for BRAKE we get

(4.43a): $\eta_{41} = \mu_{23}$ for $b_{41} = 0$ (4.45a)

(4.41a): $\eta_{42} = \min \{\mu_{11}, \mu_{21}\}$ for $b_{42} \doteq 0.25$ (4.45b)

(4.41b): $\eta_{43} = \min \{\mu_{11}, \mu_{22}\}$ for $b_{43} = 0$ (4.45c)

(4.41d): $\eta_{44} = \min \{\mu_{12}, \mu_{21}\}$ for $b_{44} = 0.25$ (4.45d)

(4.41e): $\eta_{45} = \min \{\mu_{12}, \mu_{22}\}$ for $b_{45} = 0$ (4.45e)

(4.41g): $\eta_{46} = \min \{\mu_{13}, \mu_{21}\}$ for $b_{46} = 0.75$ (4.45f)

(4.41h): $\eta_{47} = \min \{\mu_{13}, \mu_{22}\}$ for $b_{47} = 0.25$ (4.45g)

4.4.4 Defuzzification

The output conclusions from the inference engine must now be combined in a proper way to obtain a useful, continuous output signal. This is what takes place in the defuzzification interface (going from a fuzzy set to a crisp set). There are different ways of combining the outputs from the different rules. The implied membership value η_{ij} could be interpreted as the degree of certainty that the output should be b_{ij} as stated in the rule database.

The most common method is the **centre of maximum (CoM)**. This method mainly takes the weighted mean of the output membership function peak position with respect to the implied membership value produced by M rules, that is

$$ y_i = \frac{\sum_{j=1}^{M} b_{ij}\eta_{ij}}{\sum_{j=1}^{M} \eta_{ij}} \tag{4.46} $$

where y_i is the output signal, a continuous signal that can take any value between the smallest and largest output membership function peak positions. Another defuzzification method is **centre of area (CoA)** also known as the **centre of gravity (CoG)**. In this case, the respective output membership function is 'chopped off' at the level equal to the implied membership value η_{ij}. In this case, the area of the 'chopped off' (Figure 4.10) membership function A_{ij} is used as a weighting coefficient

$$ y_i = \frac{\sum_{j=1}^{M} b_{ij}A_{ij}}{\sum_{j=1}^{M} A_{ij}} \tag{4.47} $$

Yet another method of defuzzification is **mean of maximum (MoM)**. In this method, the output is chosen to be the one corresponding to the highest membership value, i.e. the b_{ij} corresponding to the largest η_{ij}. MoM is often used in managerial decision-making systems and not very often in signal processing and control systems. CoG requires more computational power than CoM, hence CoG is rarely used, while CoM is common.

Let us use CoM to complete our example. There are two continuous output signals from our cruise control, *accelerate* and *brake*. Assume *accelerate* is denoted as y_3 and *brake* as y_4. Using equation (4.46) it is straightforward to derive the defuzzification process

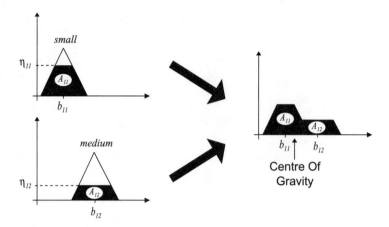

Figure 4.10 *Defuzzification using CoG*

$$y_3 = \frac{\displaystyle\sum_{j=1}^{5} b_{3j}\eta_{3j}}{\displaystyle\sum_{j=1}^{5} \eta_{3j}}$$

$$= \frac{-0.25\eta_{31} + 0.75\eta_{32} + 0.75\eta_{33} + 0.25\eta_{34} + 0.25\eta_{35}}{\eta_{31} + \eta_{32} + \eta_{33} + \eta_{34} + \eta_{35}}$$

$$= \frac{0.75(\eta_{32} + \eta_{33}) + 0.25(\eta_{34} + \eta_{35} - \eta_{31})}{\eta_{31} + \eta_{32} + \eta_{33} + \eta_{34} + \eta_{35}} \tag{4.48}$$

$$y_4 = \frac{\displaystyle\sum_{j=1}^{7} b_{4j}\eta_{4j}}{\displaystyle\sum_{j=1}^{7} \eta_{4j}}$$

$$= \frac{0\eta_{41} + 0.25\eta_{42} + 0\eta_{43} + 0.25\eta_{44} + 0\eta_{45} + 0.75\eta_{46} + 0.25\eta_{47}}{\eta_{41} + \eta_{42} + \eta_{43} + \eta_{44} + \eta_{45} + \eta_{46} + \eta_{47}}$$

$$= \frac{0.75\eta_{46} + 0.25(\eta_{42} + \eta_{44} + \eta_{47})}{\eta_{41} + \eta_{42} + \eta_{43} + \eta_{44} + \eta_{45} + \eta_{46} + \eta_{47}} \tag{4.49}$$

In Figure 4.11, plots of y_3 and y_4 are shown. It should be remembered however, that this example is simplified and that there are many alternative ways of designing fuzzy systems.

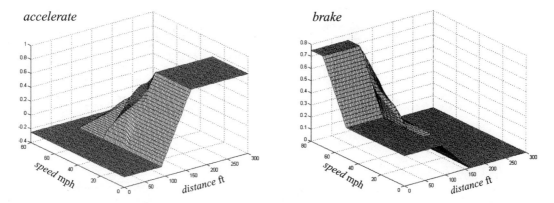

Figure 4.11 *Output signals accelerate and brake as functions of input signals speed and distance*

4.4.5 Applications

Fuzzy logic is used in some signal and image processing systems and in image identification and classifying systems. Most fuzzy systems today are however control systems. Fuzzy regulators are suitable for applications where mathematical models are hard or impossible to formulate. The process subject to control may be time varying or strongly non-linear, requiring elaborate theoretical work to be understood. Another suitable situation arises when there exists an abundant amount of practical knowledge from manual control such that experience can be formulated in natural language rather than mathematical algorithms. It is common that people having limited knowledge in control theory find fuzzy control systems easier to understand than traditional control systems.

One drawback that exists is that there are no mathematical models available and no computer simulations can be done. Numerous tests have to be performed in practice to prove the performance and stability of a fuzzy control system under all conditions. A second drawback is that the rule database has a tendency to grow large, requiring fast processing hardware to be able to perform in real time.

Fuzzy controllers can be found not only in space ships but also in air conditioners, refrigerators, microwave ovens, automatic gearboxes, cameras (auto-focus), washing machines, copying machines, distilling equipment, industrial baking processes and many other everyday applications.

5 Spectral analysis and modulation

5.1 DFT and FFT

The DFT **discrete Fourier transform** (Burrus and Parks, 1985) from the time domain to the frequency domain representation, is derived from the time-discrete Fourier transform

$$X(\omega) = \sum_{n=-\infty}^{+\infty} x(n)\, e^{-j2\pi(w/w_s)n} \tag{5.1}$$

The spectrum produced using this transform is periodic with the sampling frequency ω_s and for real input signals $x(n)$, the spectrum always has 'even' symmetry along the real axis and 'odd' symmetry on the imaginary axis. In practice, we cannot calculate this sum, since it contains an infinite number of samples. This problem is only solved by taking a section of N samples of the sequence $x(n)$. To achieve this, $x(n)$ is multiplied by a **windowing sequence** $\psi(n)$ obtaining the windowed input signal $x_N(n)$. Since multiplication in the time domain corresponds to convolution in the frequency domain, the time discrete Fourier transform of the windowed sequence will be

$$X_N(\omega) = X(\omega)^* \, \Psi(\omega) \tag{5.2}$$

where $\Psi(\omega)$ is the spectrum of the windowing sequence $\psi(n)$ and * denotes convolution. The ideal windowing sequence would have a rectangular spectrum, distorting the desired spectrum as little as possible and avoiding spectral 'leakage'. Unfortunately, a rectangular frequency response is practically impossible to build, therefore we must settle for some compromise. Commonly used windowing sequences are, for example, Rectangular, Bartlett, Hanning, Hamming and Kaiser–Bessel windows (Oppenheimer and Schafer, 1975).

Now, let us assume we have chosen an appropriate windowing sequence. Next we must determine how many frequency points should be used for calculation of the transform in order to maintain a reasonable accuracy. There is no simple answer, but in most cases good results are obtained using as many equally spaced frequency points as the number of samples in the windowed input signal, that is N. Hence the spacing between the frequency points will be ω_s/N. Now, inserting this into equation (5.1), the DFT in its most common form can be derived

$$X_N\left(k\frac{\omega_s}{N}\right) = \sum_{n=0}^{N-1} x_N(n)\, e^{-j2\pi(kn/N)} = \sum_{n=0}^{N-1} x_N(n)\, W_N^{kn} \tag{5.3}$$

where the **twiddle factor** is

$$W_N = e^{-j(2\pi/N)} \qquad (5.4)$$

Unfortunately, the number of complex computations needed to perform the DFT is proportional to N^2. The acronym FFT (**Fast Fourier Transform**) refers to a group of algorithms, all very similar, which uses fewer computational steps to efficiently compute the DFT. The number of steps are typically proportional to $N\text{lb}(N)$, where $\text{lb}(x) = \log_2(x)$ is the logarithm base 2. Reducing the number of computational steps is of course important if the transform has to be computed in a real time system. Fewer steps implies faster processing time, hence higher sampling rates are possible. Now, there are essentially two tricks employed to obtain this 'sped up version' of DFT.

(1) When calculating the sum (5.3) for $k = 0, 1, 2 \ldots N$, many complex multiplications are repeated. By doing the calculations in a 'smarter' order, many calculated complex products can be stored and reused.
(2) Further, the twiddle factor (5.4) is periodic and only N factors need to be computed. This can of course be done in advance and the values can be stored in a table.

Let us illustrate the ideas behind the FFT algorithm by using a simple example having $N = 4$. If we use the original DFT transform as is equation (5.3), the following computations are needed

$$k=0: \; X_4(0) = x_4(0)W_4^0 + x_4(1)W_4^0 + x_4(2)W_4^0 + x_4(3)W_4^0$$

$$k=1: \; X_4(1) = x_4(0)W_4^0 + x_4(1)W_4^1 + x_4(2)W_4^2 + x_4(3)W_4^3$$

$$k=2: \; X_4(2) = x_4(0)W_4^0 + x_4(1)W_4^2 + x_4(2)W_4^4 + x_4(3)W_4^6 \qquad (5.5)$$

$$k=3: \; X_4(3) = x_4(0)W_4^0 + x_4(1)W_4^3 + x_4(2)W_4^6 + x_4(3)W_4^9$$

As can be seen from computations (5.5), 16 complex multiplications and 12 complex additions are needed. Now, let us see how these numbers may be reduced. Firstly, if we put the odd and even numbered terms in two groups and divide the odd terms by W_4^k computations (5.5) can be rewritten as

$$X_4(0) = \left(x_4(0)W_4^0 + x_4(2)W_4^0\right) + W_4^0\left(x_4(1) + x_4(3)\right)$$

$$X_4(1) = \left(x_4(0)W_4^0 + x_4(2)W_4^2\right) + W_4^1\left(x_4(1) + x_4(3)W_4^2\right)$$

$$X_4(2) = \left(x_4(0)W_4^0 + x_4(2)W_4^4\right) + W_4^2\left(x_4(1) + x_4(3)W_4^4\right) \qquad (5.6)$$

$$X_4(3) = \left(x_4(0)W_4^0 + x_4(2)W_4^6\right) + W_4^3\left(x_4(1) + x_4(3)W_4^6\right)$$

Secondly, we use the periodicity of the twiddle factor (5.4)

$$W_4^4 = W_4^0, \;\; W_4^6 = W_4^2$$

Further, we know that $W_4^0 = 1$. Inserting this into computations (5.6), we obtain

$$X_4(0) = (x_4(0) + x_4(2)) + (x_4(1) + x_4(3)) = A + C$$

$$X_4(1) = (x_4(0) + x_4(2)W_4^2) + W_4^1(x_4(1) + x_4(3)W_4^2) = B + W_4^1 D$$

$$X_4(2) = (x_4(0) + x_4(2)) + W_4^2(x_4(1) + x_4(3)) = A + W_4^2 C \qquad (5.7)$$

$$X_4(3) = (x_4(0) + x_4(2)W_4^2) + W_4^3(x_4(1) + x_4(3)W_4^2) = B + W_4^3 D$$

From computations (5.7) we can now see that our computations can be executed in two steps. In step one we calculate the terms A, B, C and D and in step two we calculate the transform values $X_4(k)$. Hence step 1 is

$$
\begin{aligned}
A &= x_4(0) + x_4(2) \\
B &= x_4(0) + x_4(2)W_4^2 = x_4(0) - x_4(2) \\
C &= x_4(1) + x_4(3) \\
D &= x_4(1) + x_4(3)W_4^2 = x_4(1) - x_4(3)
\end{aligned}
\qquad (5.8)
$$

This step requires two complex additions and two complex subtractions. If all input signals are real, we only need real additions and subtractions. Step 2 is

$$
\begin{aligned}
X_4(0) &= A + C \\
X_4(1) &= B + W_4^1 D \\
X_4(2) &= A + W_4^2 C = A - C \\
X_4(3) &= B + W_4^3 D = B - W_4^1 D
\end{aligned}
\qquad (5.9)
$$

Two complex multiplications, two complex additions and two subtractions are needed. In total, we now have two complex multiplications and eight complex additions (subtractions). This is a considerable saving compared to the 16 multiplications and 12 additions required to compute the DFT in its original form.

The steps of the FFT (in our example steps (5.8) and (5.9)) are often described in signal flow chart form, denoted 'FFT butterflies' (see Figure 5.1). For the general case, the FFT strategy can be expressed as

$$X_N\left(k\frac{\omega_s}{N}\right) = \sum_{n=0}^{N-1} x_N(n)W_N^{kn} =$$

$$\sum_{n=0}^{N/2-1} x_N(2n)W_{N/2}^{kn} + W_N^k \sum_{n=0}^{N/2-1} x_N(2n+1)W_{N/2}^{kn} \qquad (5.10)$$

and

$$W_N^{kn} = W_N^{(kn+jN)} \quad \text{where } j = 1, 2, \ldots \qquad (5.11)$$

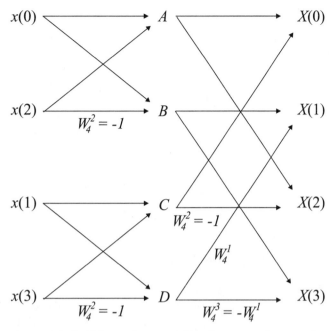

Figure 5.1 *FFT butterfly signal flow diagram, showing the example having N = 4*

5.2 Spectral analysis

Spectral analysis, by estimation of the **power spectrum** or **spectral power density** of a deterministic or random signal, involves performing a squaring function. Obtaining a good estimate of the spectrum, that is the signal power contents as a function of the frequency, is not entirely easy in practice. The main problem is that in most cases we only have access to a limited set of samples of the signal; or in another situation, we are forced to limit the number of samples to be able to perform the calculations in a reasonable time. These limitations introduce errors into the estimate. If the signal is a random type signal, we may also obtain large fluctuations in power estimates based on samples from different populations. Hence, in the general case there is no way to obtain the **true** power spectrum of a signal unless we are allowed to observe it infinitely. This is why the term **estimation** is frequently used in this context.

5.2.1 DFT and FFT approaches

Spectral analysis using the Fourier transform, a non-parametric method, was originally proposed in 1807 by the French mathematician and Baron J.B.J. Fourier. The discrete version of this transform, commonly used today in digital signal processing applications, is called **Discrete Fourier Transform (DFT)**. A smart algorithm for calculating the DFT, causing less computational load for a digital computer, is the **Fast Fourier Transform (FFT)**. From a purely mathematical point of view, DFT and FFT do the same job as shown above. Consider a sampled signal $x(n)$ for $-\infty < n < \infty$ further and assume that the signal has finite energy, that is

$$\sum_{n=-\infty}^{\infty} x^2(n) < \infty \qquad (5.12)$$

The DFT of the signal can then be expressed as

$$X(\omega) = \sum_{n=-\infty}^{\infty} x(n)\, e^{-j2\pi(w/w_s)n} \qquad (5.13)$$

where ω_s is the sampling frequency. By Parseval's relation, the energy of the signal can then be expressed as

$$\sum_{n=-\infty}^{\infty} |x(n)|^2 = \frac{1}{2\pi}\int_{-\pi}^{\pi} |X(\omega)|^2\, d\omega = \frac{1}{2\pi}\int_{-\pi}^{\pi} S_{xx}(\omega)\, d\omega \qquad (5.14)$$

where $S_{xx}(\omega)$ is referred to as the **spectral density** of the signal $x(n)$. The spectral density is hence the squared magnitude of the Fourier transform of the signal. As mentioned above, in most cases the signal is not defined for $-\infty < n < \infty$, or we may not be able to handle an infinite length sequence in a practical application. In such a case, the signal is assumed to be non-zero only for $n = 0, 1, \ldots N-1$ and assumed to be zero otherwise. The spectral density is then written as

$$S_{xx}(\omega) = |X(\omega)|^2 = \left| \sum_{n=0}^{N-1} x(n)\, e^{-j2\pi(\omega/\omega_s)n} \right|^2 \qquad (5.15)$$

The power of the signal as a function of the frequency is called a **periodogram**, A. Schuster defined it in 1898 as

$$P(\omega) = \frac{1}{N} S_{xx}(\omega) \qquad (5.16)$$

In the case where the signal $x(n)$ exists over the entire interval $(-\infty, \infty)$ and its energy is infinite (e.g. sinusoidal or other periodic signal) it is convenient to define the spectral density as (Mitra and Kaiser, 1993)

$$\Phi_{xx}(\omega) = \lim_{N\to\infty} \frac{1}{2N+1} \left| \sum_{n=-N}^{N} x(n)\, e^{-j2\pi(\omega/\omega_s)n} \right|^2 \qquad (5.17)$$

Now, in many practical cases, to use an FFT algorithm, N is chosen to be a power of 2 (e.g. 64, 128, 256 . . .). This means that the spectral density can only be calculated at N discrete points, which in many cases turns out to be too sparse. Quite often, we need a finer frequency spacing, i.e. we need to know the spectrum at L points where $L > N$. This can be accomplished by **zero-padding**, so that the data sequence consisting of N samples, is extended by adding $L - N$ zero value samples to the end of the sequence. The 'new' L-point sequence is then transformed using DFT or FFT. The effective frequency spacing is now

$$\frac{2\pi}{L\omega_s} < \frac{2\pi}{N\omega_s} \qquad (5.18)$$

It is important to remember that zero-padding **does not** increase the resolution in frequency, it merely **interpolates** the spectrum at more points. For some applications this is however sufficient. So far, we have assumed that the signal $x(n)$ has been identical to zero outside the interval $n = 0, 1, \ldots N-1$. This is equivalent to multiplying the sequence $x(n)$ with a rectangular window sequence

$$s(n) = x(n)w(n) \qquad (5.19)$$

where the **rectangular window** can be expressed as

$$w(n) = \begin{cases} 1 & \text{for } 0 \leqslant n < N \\ 0 & \text{else} \end{cases} \qquad (5.20)$$

We recall from basic signal theory that a multiplication in the time domain, like equation (5.19) corresponds to a convolution in the frequency domain. Therefore, the result of the windowing process has the 'true' spectrum of the signal convoluted with the Fourier transform of the window sequence. What we obtain is not the spectrum of the signal, but rather the spectrum of the signal $x(n)$ **distorted** by the transform of the window sequence. This distortion results in 'smearing' of the spectrum, which implies that narrow spectral peaks cannot be detected nor distinguished from each other (Mitra and Kaiser, 1993; Lynn and Fuerst, 1994). This 'leakage' is an inherent limitation when using conventional Fourier techniques for spectral analysis. The only way to reduce this effect is to observe the signal for a longer duration, i.e. gather more samples. The quality of the spectral estimate can however be somewhat improved by using a 'smoother' window sequence than the quite 'crude' rectangular window. Some examples of windows with different distortion properties (Oppenheimer and Schafer, 1975; Mitra and Kaiser, 1993) are given below.

The Bartlett window (triangular window)

$$w(n) = \begin{cases} 2n/(N-1) & 0 \leqslant n < (N-1)/2 \\ 2 - 2n/(N-1) & (N-1)/2 \leqslant n < N \end{cases} \qquad (5.21)$$

The Hann window

$$w(n) = \frac{1}{2}\left(1 - \cos\left(\frac{2n\pi}{N-1}\right)\right) \quad 0 \leqslant n < N \qquad (5.22)$$

The Hamming window

$$w(n) = 0.54 - 0.46 \cos\left(\frac{2n\pi}{N-1}\right) \quad 0 \leqslant n < N \qquad (5.23)$$

The Blackman window

$$w(n) = 0.42 - 0.5 \cos\left(\frac{2n\pi}{N-1}\right) + 0.08 \cos\left(\frac{4n\pi}{N-1}\right) \quad 0 \leqslant n < N \qquad (5.24)$$

5.2.2 Using the auto-correlation function

In the previous section, we concluded that a spectral power estimate could be obtained by taking the square of the magnitude of the Fourier transform (5.15) of the signal $x(n)$. There is an alternative way of achieving the same result by using the **auto-correlation function** of the signal $x(n)$. Let us start this section by a brief discussion on correlation functions (Papoulis, 1985; Denbigh, 1998; Schwartz and Shaw, 1975) of time-discrete signals. Assume that the means of the two complex signal sequences $x(n)$ and $y(n)$ are zero

$$E[x(n)] = \eta_x = 0 \quad \text{and} \quad E[y(n)] = \eta_y = 0$$

The **cross-correlation** between these signals is then defined as

$$R_{xy}(n, m) = E[x(n)y^*(m)] \qquad (5.25)$$

where * denotes a complex conjugate and $E[\ ]$ is the expected mean operator (ensemble average). If the mean of the signals is not zero, the **cross-covariance** is

$$C_{xy}(n, m) = R_{xy}(n, m) - \eta_x(n)\eta_y^*(m) \qquad (5.26)$$

Note! if the mean is zero, the cross-correlation and cross-covariance are equal. If the signal is correlated by itself, **auto-correlation** results

$$R_{xx}(n, m) = E[x(n)x^*(m)] \qquad (5.27)$$

As a consequence of the above, the **auto-covariance** is defined as

$$C_{xx}(n, m) = R_{xx}(n, m) - \eta_x(n)\eta_x^*(m) \qquad (5.28)$$

If the signal $x(n)$ has zero mean and the auto-correlation only depends on the **difference** $k = m-n$ and not on the actual values of n or m themselves, the signal $x(n)$ is said to be **wide-sense stationary** from a statistical point of view. This basically means that regardless of which point in the sequence the statistical properties of the signal is studied, they all turn out to be similar. For a stationary signal (which is a common assumption on signals) the **auto-correlation** function is then defined as

$$R_{xx}(k) = E[x(n)x^*(n+k)] \qquad (5.29)$$

Note in particular that $R_{xx}(0) = \sigma^2$ is the **variance** of the signal and represents the power of the signal. Now, starting from equation (5.17), we study the square magnitude of the Fourier-transformed signal sequence $x(n)$ as $N \rightarrow \infty$

$$\left| \sum_{n=-\infty}^{\infty} x(n)\, e^{-j2\pi(\omega/\omega_s)n} \right|^2 = \left(\sum_{n=-\infty}^{\infty} x(n)\, e^{-j2\pi(\omega/\omega_s)n} \right) \left(\sum_{n=-\infty}^{\infty} x(n)\, e^{-j2\pi(\omega/\omega_s)n} \right)^*$$

$$= \sum_{n=-\infty}^{\infty} \sum_{m=-\infty}^{\infty} x(n)x^*(m)\, e^{-j2\pi(\omega/\omega_s)(m-n)}$$

$$= \sum_{n=-\infty}^{\infty} \sum_{k=-\infty}^{\infty} x(n)x^*(n+k)\, e^{-j2\pi(\omega/\omega_s)k}$$

$$= \sum_{n=-\infty}^{\infty} R_{xx}(k)\, e^{-j2\pi(\omega/\omega_s)k} \qquad (5.30)$$

Hence, we now realize that the spectral density can be obtained in an **alternative way**, namely by taking the DFT of the auto-correlation function

$$\Phi_{xx}(\omega) = \sum_{n=-\infty}^{\infty} R_{xx}(k)\, e^{-j2\pi(\omega/\omega_s)k} \qquad (5.31)$$

This relationship is known as the Wiener–Khintchine theorem.

5.2.3 Periodogram averaging

When trying to obtain spectral power estimates of stochastic signals using a limited number of samples, quite poor estimates often result. This is especially true near the ends of the sample window, where the calculations are based on few samples of the signal $x(n)$. These poor estimates cause wild fluctuations in the periodogram. This trend had already been observed by A. Schuster in 1898. To get smoother power spectral density estimates, many independent periodograms have to be averaged. This was first studied by M.S. Bartlett and later by P.D. Welch.

Assume that $x(n)$ is a stochastic signal being observed in L points, i.e. for $n = 0, 1, \ldots L-1$. This is the same as multiplying the signal $x(n)$ by a rectangular window $w(n)$, being non-zero for $n = 0, 1, \ldots L-1$ (see expression (5.20)). Using the DFT (5.13) in this product we obtain

$$S(\omega) = \sum_{n=0}^{L-1} x(n)w(n)\, e^{-j2\pi(\omega/\omega_s)n} \qquad (5.32)$$

We now consider an estimate of the power spectral density (Mitra and Kaiser, 1993) given by

$$\hat{I}_{xx}(\omega) = \frac{1}{LU}\, |S(\omega)|^2 \qquad (5.33)$$

where U is a normalizing factor to remove any bias caused by the window $w(n)$

$$U = \frac{1}{L} \sum_{n=0}^{L-1} |w(n)|^2 \qquad (5.34)$$

The entity $\hat{I}_{xx}(\omega)$ is denoted as a **periodogram** if the window used is the rectangular window, else it is called a **modified periodogram**. Now, **Welch's method** computes periodograms of overlapping segments of data and averages them (Mitra and Kaiser, 1993). Assume that we have access to Q contiguous samples of $x(n)$ and that $Q > L$. We divide the data set into P segments of L samples each. Let S be the shift between successive segments, then

$$Q = (P-1)S + L \tag{5.35}$$

Hence, the 'windowed' segment p is then expressed as

$$s^{(p)}(n) = w(n)x(n + pS) \tag{5.36}$$

for $0 \leqslant n < L$ and $0 \leqslant p < P$. Inserting equation (5.35) into equation (5.32) we obtain the DFT of segment p

$$S^{(p)}(\omega) = \sum_{n=0}^{L-1} s^{(p)}(n) \, e^{-j2\pi(w/w_s)n} \tag{5.37}$$

Using equation (5.33), the periodogram of segment p can be calculated

$$\hat{I}_{xx}^{(p)}(\omega) = \frac{1}{LU} |S^{(p)}(\omega)|^2 \tag{5.38}$$

Finally, the **Welch estimate** is the average of all the periodograms over all segments p as above

$$\hat{I}_{xx}^{W}(\omega) = \frac{1}{P} \sum_{p=0}^{P-1} \hat{I}_{xx}^{(p)}(\omega) \tag{5.39}$$

For the Welch estimate, the bias and variance asymptotically approach zero as Q increases. For a given Q, we should choose L as large as possible to obtain the best resolution, but on the other hand, to obtain a smooth estimate, P should be large, implying L to be small (equation (5.35)). Hence, there is a **tradeoff** between high resolution in frequency (large L) and smooth spectral density estimate (small L). Increasing Q is of course always good, but requires longer data acquisition time and more computational power.

The **Bartlett periodogram** is a special case of Welch's method as described above. In Bartlett's method, the segments are **non-overlapping**, that is, $S = L$ in equation (5.35). The same tradeoff described above for Welch's method also applies to Bartlett's method. In terms of variance in the estimate, Welch's method often performs better than Bartlett's method. Implementing Bartlett's method may however in some cases be somewhat easier in practice.

5.2.4 Parametric spectrum analysis

Non-parametric spectral density estimation methods like Fourier analysis as described above are well studied and established. Unfortunately, this class of methods has some drawbacks. When the available set of N samples is small, resolution in frequency is severely limited. Also auto-correlation outside the sample set is considered to be zero, i.e. $R_{xx}(k) = 0$ for $k > N$, which may be an unrealistic assumption in many cases.

Fourier-based methods assume that data outside the observed window are either periodic or zero. The estimate is not only an estimate of the observed N data samples, but also an estimate of the 'unknown' data samples outside the window, under the assumptions above. Alternative estimation methods can be found in the class of **model-based, parametric spectrum analysis methods** (Mitra and Kaiser, 1993). Some of the most common methods will be presented briefly below.

The underlying idea of assuming the signal to be analysed can be generated using a **model filter**. The filter has the causal transfer function $H(z)$ and a white noise input signal $e(n)$, with mean value zero and variance σ_e^2. In this case, the output signal $x(n)$ from the filter is a wide-sense stationary signal with power density

$$\Phi_{xx}(\omega) = \Phi_e(\omega)|H(\omega)|^2 = \sigma_e^2|H(\omega)|^2 \qquad (5.40)$$

Hence, the power density can be obtained if the model filter transfer function $H(z)$ is known, i.e. if the model type and its associated filter **parameters** are known. The parametric spectrum analysis can be divided into three steps:

- selecting an appropriate model and selecting the order of $H(z)$. There are often many different possibilities
- estimating the filter coefficients, i.e. the parameters from the N data samples $x(n)$ where $n = 0, 1, \ldots N-1$
- evaluating the power density, as in equation (5.40) above.

Selecting a model is often easier if some prior knowledge of the signal is available. Different models may give more or less accurate results, but may also be more or less computationally demanding. A common model type is the **ARMA (auto-regressive moving average)** model

$$H(z) = \frac{B(z)}{A(z)} = \frac{\displaystyle\sum_{k=0}^{q} b_k z^{-k}}{1 + \displaystyle\sum_{k=1}^{p} a_k z^{-k}} = \sum_{n=0}^{\infty} h(n)z^{-n} \qquad (5.41)$$

where $h(n)$ is the impulse response of the filter and the denominator polynomial $A(z)$ has all its roots inside the unit circle for the filter to be stable. The ARMA model may be simplified. If $q = 0$, then $B(z) = 1$ and an all-pole filter results, yielding the **AR (auto-regressive)** model

$$H(z) = \frac{1}{1 + \displaystyle\sum_{k=1}^{p} a_k z^{-k}} \qquad (5.42)$$

or, if $p = 0$, then $A(z) = 1$ and a filter having only zeros results, a so-called **MA (moving average)** model

$$H(z) = \sum_{k=0}^{q} b_k z^{-k} \qquad (5.43)$$

Note that these models are mainly IIR and FIR filters (see also Chapter 1). Once a reasonable model is chosen, the next measure is estimating the filter parameters based on the available data samples. This can be done in a number of ways. One way is using the auto-correlation properties of the data sequence $x(n)$. Assuming an ARMA model as in equation (5.41), the corresponding difference equation (see Chapter 1) can be written

$$x(n) = -\sum_{k=1}^{p} a_k x(n-k) + \sum_{k=0}^{q} b_k e(n-k) \tag{5.44}$$

where $e(n)$ is the white noise input sequence. Multiplying equation (5.44) by $x^*(n + m)$ and taking the expected value we obtain

$$R_{xx}(m) = -\sum_{k=1}^{p} a_k R_{xx}(m-k) + \sum_{k=0}^{q} b_k R_{ex}(m-k) \tag{5.45}$$

where the auto-correlation of $x(n)$ is

$$R_{ex}(m) = E[x(n)x^*(n+m)] \tag{5.46}$$

and the cross-correlation between $x(n)$ and $e(n)$ is

$$R_{ex}(m) = E[e(n)x^*(n+m)] \tag{5.47}$$

Since $x(n)$ is the convolution of $h(n)$ by $e(n)$, $x^*(n + m)$ can be expressed as

$$x^*(n+m) = \sum_{k=0}^{\infty} h^*(k)\, e(n+m-k) \tag{5.48}$$

Inserting into equation (5.47), the cross-correlation can hence be rewritten as

$$R_{ex}(m) = E\left[e(n) \sum_{k=0}^{\infty} h^*(k)\, e(n+m-k) \right]$$

$$= \sum_{k=0}^{\infty} h^*(k)\, E[e(n)e(n+m-k)] = \sigma_e^2 h^*(-m) \tag{5.49}$$

where we have used the facts that $e(n)$ is a white noise signal and $h(n)$ is causal, i.e. zero for $n < 0$. We can now express the auto-correlation values for the signal originating from the ARMA model, in terms of the model parameters

$$R_{xx}(m) = \begin{cases} -\sum_{k=1}^{p} a_k R_{xx}(m-k) & \text{for } m \geqslant q \\[2ex] -\sum_{k=1}^{p} a_k R_{xx}(m-k) + \sigma_e^2 \sum_{k=0}^{q-m} h^*(k) b_{k+m} & \text{for } 0 \leqslant m < q \\[2ex] R_{xx}^*(-m) & \text{for } m < 0 \end{cases} \tag{5.50}$$

The auto-correlation values $R_{xx}(m)$ for $|m| \geqslant q$ are extrapolated by the filter parameters and the values of $R_{xx}(m)$ for $m = 0, 1, \ldots q$. Now, for the sake of simplicity, assume that we are using an AR model, i.e. $q = 0$. Then equation (5.50) simplifies to

$$R_{xx}(m) = \begin{cases} -\sum_{k=1}^{p} a_k R_{xx}(m-k) & \text{for } m \geqslant 0 \\[2em] -\sum_{k=1}^{p} a_k R_{xx}(m-k) + \sigma_e^2 & \text{for } 0 = m \\[2em] R_{xx}^*(-m) & \text{for } m < 0 \end{cases} \quad (5.51)$$

Thus, if the auto-correlation values $R_{xx}(0)$, $R_{xx}(1)$, ... $R_{xx}(p)$ are known, the filter parameters can be found by solving p linear equations corresponding to $m = 0, 1, ... p$. Note that we only need to know $p+1$ correlation values to be able to determine all the parameters. From equation (5.51) and setting $m = 0$ we can also obtain the variance

$$\sigma_e^2 = R_{xx}(0) + \sum_{k=1}^{p} a_k R_{xx}(-k) \quad (5.52)$$

Combining equations (5.51) and (5.52) we thus have to solve the following equations to obtain the parameters. These equations are known as the **Yule–Walker equations**.

$$R_{xx}(0) + a_1 R_{xx}(-1) + ... a_p R_{xx}(-p) = \sigma_e^2$$

$$R_{xx}(1) + a_1 R_{xx}(0) + ... a_p R_{xx}(-p+1) = 0$$

$$\vdots \qquad\qquad \vdots \qquad\qquad \vdots \qquad\qquad (5.53)$$

$$R_{xx}(p) + a_1 R_{xx}(p-1) + ... a_p R_{xx}(0) = 0$$

For the ARMA model, the modified Yule–Walker equations can be used to determine the filter parameters. The Yule–Walker equations can be solved using standard techniques, for instance Gauss elimination, which requires a number of operations proportional to p^3. There are however recursive, smarter algorithms that make use of the regular structure of the Yule–Walker equations and hence require fewer operations. One such algorithm is the **Levinson–Durbin algorithm** (Mitra and Kaiser, 1993) requiring a number of operations proportional to p^2 only. There are also other ways to obtain the parameters. One such way is **adaptive modelling** using adaptive filters (Widrow and Stearns, 1985) (see Chapter 3).

Finally, when the filter parameters are estimated, the last step is to evaluate the spectral density. This may be achieved by inserting

$$z = e^{-j2\pi(w/w_s)} \quad (5.54)$$

into the transfer function $H(z)$ of the model (for example equation (5.41)) and then, using the relation (5.40), evaluating the spectral density. It should however be pointed out that the calculations needed may be tiring and that considerable error between the estimated and the true spectral density may occur. One has to remember that the spectral density is an approximation, based on the model used. Choosing an improper model for a specific signal may hence result in a poor spectral estimate.

The parametric analysis methods have other interesting features. In some applications, the parameters themselves are sufficient information about the signal. In speech-coding equipment for instance, the parameters are transmitted to the receiver instead of the speech signal itself. Using a model filter and a noise source at the receiving site, the speech signal can be recreated. This technique is a type of data compression algorithm (see also Chapter 7), since transmitting the filter parameters requires less capacity then transmitting the signal itself.

5.2.5 Wavelet analysis

Wavelet analysis, also called **wavelet theory** or just **wavelets** (Lee and Yamamoto, 1994; Bergh *et al.*, (1999) has attracted much attention recently. It has been used in transient analysis, image analysis and communication systems and has shown to be very useful for processing **non-stationary** signals.

Wavelet analysis deals with expansion of functions in terms of a set of **basis functions**, like Fourier analysis. Instead of trigonometric functions being used, as in Fourier analysis, wavelets are used. The objective of wavelet analysis is to define these wavelet basis functions. It can be shown that every application using the FFT can be reformulated using wavelets. In the latter case, more localized temporal and frequency information can be provided. Thus, instead of a conventional frequency spectrum, a 'wavelet spectrum' may be obtained.

Wavelets are created by **translations** and **dilations** of a fixed function called the **mother wavelet**. Assume that $\psi(t)$ is a complex function. If this function satisfies the following two conditions, it may be used as a mother wavelet

Condition one

$$\int_{-\infty}^{\infty} |\psi(t)|^2 \, dt < \infty \tag{5.55}$$

This expression implies finite energy of the function.

Condition two

$$c_\psi = 2\pi \int_{-\infty}^{\infty} \frac{|\Psi(\omega)|^2}{\omega} \, d\omega < \infty \tag{5.56}$$

where $\Psi(\omega)$ is the Fourier transform of $\psi(t)$. This condition implies that if $\Psi(\omega)$ is smooth, then $\Psi(0) = 0$. One example of a mother wavelet is the **Haar** wavelet, which is the historically original and simplest wavelet

$$\psi(t) = \begin{cases} 1 & \text{for } 0 < t \leq 1/2 \\ -1 & \text{for } 1/2 < t \leq 1 \\ 0 & \text{else} \end{cases} \tag{5.57}$$

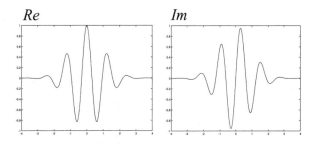

Figure 5.2 *Real and imaginary part of a **Morlet** wavelet*

Other examples of smooth wavelets having better frequency localization properties are the **Meyer** wavelet, the **Morlet** wavelet and the **Daubechies** wavelet (Lee and Yamamoto, 1994; Bergh *et al.*, 1999). Figure 5.2 shows an example of a Morlet wavelet. The wavelets are obtained by translations and dilations of the mother wavelet

$$\psi_{a,b}(t) = a^{-1/2}\psi\left(\frac{t-b}{a}\right) \tag{5.58}$$

which means rescaling the mother wavelet by a and shifting in time by b, where $a > 0$ and $-\infty < b < \infty$.

The **wavelet transform** of the real signal $x(t)$ is the scalar product

$$X(b, a) = \int_{-\infty}^{\infty} \psi_{a,b}^*(t)\, x(t)\, dt \tag{5.59}$$

where * denotes the complex conjugate. There is also an inverse transform

$$x(t) = \frac{1}{c_\psi}\int_{-\infty}^{\infty}\int_0^{\infty} X(b, a)\psi_{a,b}(t)\,\frac{da\, db}{a^2} \tag{5.60}$$

where c_ψ is obtained from equation (5.56). For the **discrete wavelet transform**, equations (5.58), (5.59) and (5.60) are replaced by equations (5.61), (5.62) and (5.63) respectively

$$\psi_{m,n}(t) = a_0^{-m/2}\,\psi\left(\frac{t-nb_0}{a_0^m}\right) \tag{5.61}$$

where m and n are integers and a_0 and b_0 are constants. Quite often a_0 is chosen as

$$a_0 = 2^{1/v}$$

where v is an integer and v pieces of $\psi_{m,n}(t)$ are processed as one group called a **voice**

$$X_{m,n} = \int_{-\infty}^{\infty} \psi_{m,n}^*(t)x(t)\, dt \tag{5.62}$$

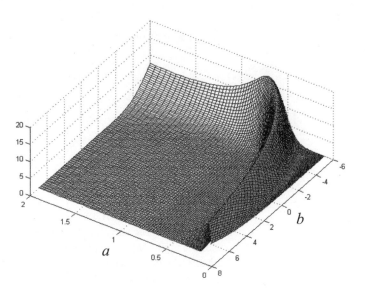

Figure 5.3 *Magnitude of a wavelet transform (Morlet) of a sinusoid with linearly increasing frequency*

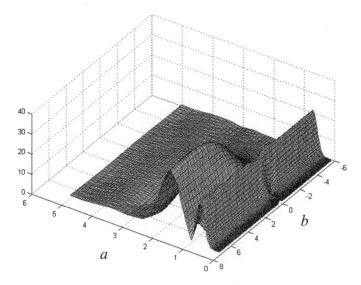

Figure 5.4 *A signal comprising two sinusoids with different frequencies, where one frequency component is switched on with a time delay*

$$x(t) = k_\psi \sum_m \sum_n X_{m,n}\, \psi_{m,n}(t) \qquad (5.63)$$

Graphical representation of the complex functions obtained from equations (5.59) and (5.62) is not always entirely easy. Commonly, 3D graphics or grey-scale graphics showing the magnitude and phase as a function of a and b, are used. One example is Figure 5.3, showing the magnitude of a wavelet transform (Morlet) of a sinusoid with linearly increasing frequency. Figure 5.4 shows a signal comprising two sinusoids with different frequencies, where one frequency component is switched on with a time delay.

5.3 Complex modulation

In many communication systems, especially in radio systems, the information ('digital' or 'analog') to be transmitted is **modulated**, i.e. 'encoded' onto a **carrier** signal. The carrier is chosen depending on the type of media available for the transmission, for instance, a frequency band in the radio spectrum or an appropriate frequency for transmission over cables or fibres. If we assume that the carrier signal is purely a **cosine** signal, there are three parameters of the carrier that can be modulated that are being used for information transfer. These parameters are the **amplitude** $a(t)$, the **frequency** $f_c(t)$ and the **phase** $\phi(t)$. The modulated carrier is then

$$s(t) = a(t)\cos\big(2\pi f_c(t)t + \phi(t)\big) \qquad (5.64)$$

The corresponding modulation types are denoted **amplitude modulation (AM), frequency modulation (FM)** and **phase modulation (PM)**. Note that FM and PM are related in that the frequency is the phase changing speed, in other words, the derivative of the phase function is the frequency. Hence, FM can be achieved by changing the phase in a proper way.

In a **digital** radio communication system it is common to modulate either by changing $f_c(t)$ between a number of discrete frequencies, **FSK (frequency shift keying)** or by changing $\phi(t)$ between a number of discrete phases, **PSK (phase shift keying)**. In some systems, the amplitude is also changed between discrete levels **ASK (amplitude shift keying)**. So if we are to design a modulation scheme being able to transmit M different discrete symbols (for binary signals, $M = 2$), we hence have to define M unique combinations of amplitude, frequency and/or phase values of the carrier signal. The trick is to define these **signal points** (triplets of $a(t)$, $f_c(t)$ and $\phi(t)$) in a way that communication speed can be high, influence of interference low, spectral occupancy low and modulation and demodulation equipment made fairly simple.

A good approach when working with modulation is to utilize the concept of **complex modulation**, a general method being able to achieve all the modulation types mentioned above.

5.3.1 Complex representation of narrowband signals

Assume that the bandwidth of the **baseband signal**, the information modulated onto the carrier, is small relative to the frequency of the carrier signal. This implies that the modulated carrier, the signal being transmitted over the communications media, can be regarded a **narrowband passband signal**, or simply a **bandpass signal** (Proakis, 1989; Ahlin and Zander, 1998)

$$s(t) = a(t) \cos \left(2\pi f_c t + \phi(t)\right)$$

$$= a(t) \cos (\phi(t)) \cos (2\pi f_c t) - a(t) \sin (\phi(t)) \sin (2\pi f_c t)$$

$$= x(t) \cos (2\pi f_c t) - y(t) \sin (2\pi f_c t) \tag{5.65}$$

where $x(t)$ and $y(t)$ are the **quadrature components** of the signal $s(t)$, defined as

$$x(t) = a(t) \cos (\phi(t)) \quad \text{the \textbf{in phase} or } I \text{ component} \tag{5.66a}$$

and $\quad y(t) = a(t) \sin (\phi(t)) \quad$ the **quadrature phase** or Q component (5.66b)

The two components above can be assembled into one handy entity by defining the **complex envelope**

$$z(t) = x(t) + jy(t) = a(t) \cos (\phi(t)) + ja(t) \sin (\phi(t)) = a(t) \, e^{j\phi(t)} \tag{5.67}$$

Hence, the modulated carrier can now be written (see also equation (5.65)) as

$$s(t) = \text{Re} \left[z(t) \, e^{j2\pi f_c t} \right] \tag{5.68}$$

The modulation process is performed by simply making a **complex multiplication** of the carrier and the complex envelope. This can be shown on component level

$$s(t) = x(t) \cos (2\pi f_c t) - y(t) \sin (2\pi f_c t) \tag{5.69}$$

Figure 5.5 shows a simple, digital quadrature modulator implementing expression (5.69) above. The *M*-ary symbol sequence to be transmitted enters the

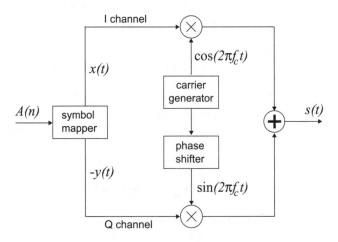

Figure 5.5 *Simple M-ary quadrature modulator*

Table 5.1 *Example: I and Q components and phase angle as a function of input bits*

$A(n)$	$A(n-1)$	$\phi(t)$	$x(t)$	$y(t)$
0	0	45°	$1/\sqrt{2}$	$1/\sqrt{2}$
0	1	135°	$-1/\sqrt{2}$	$1/\sqrt{2}$
1	0	−45°	$1/\sqrt{2}$	$-1/\sqrt{2}$
1	1	−135°	$-1/\sqrt{2}$	$-1/\sqrt{2}$

symbol mapper. This device determines the quadrature components $x(t)$ and $y(t)$ according to the modulation scheme in use.

Let us for example assume that the incoming data stream $A(n)$ consists of binary digits, that is, ones and zeros. We have decided to use a four-phase PSK (QPSK), in other words we are using $M = 4$ symbols represented by four equally spaced phase shifts of the carrier signal. So, two binary digits $A(n)$ and $A(n-1)$ are transmitted simultaneously. If we assume that $a(t) = 1$ (constant amplitude), the mapper hence implements a table like Table 5.1 (many schemes exist).

The quadrature components coming from the mapper, having half the data rate of the binary symbols, are then multiplied by the carrier and a $-90°$ phase-shifted version (the quadrature channel) of the carrier, since $\cos(\alpha - 90) = \sin(\alpha)$. The I and Q channels are finally added and the modulated carrier $s(t)$ is obtained.

Demodulation of the signal at the receiving site can easily be done in a similar way, by performing a complex multiplication

$$\hat{z}(t) = s(t)\,e^{-j2\pi f_c(t)} \tag{5.70}$$

and removing high-frequency components using a low-pass filter. From the received estimate $\hat{z}(t)$ the quadrature components can be obtained and from these the information symbols. In practice, the situation is more complicated, because the received signal is not $s(t)$, but rather a distorted version, with noise and interference added.

5.3.2 The Hilbert transformer

A special kind of 'ideal' filter is the **quadrature filter**. Since this filter is non-causal, only approximations (Mitra and Kaiser, 1993) of the filter can be implemented in practice. This filter model is often used when dealing with **single sideband (SSB)** signals. The ideal frequency response is

$$H(\omega) = \begin{cases} -j & \omega > 0 \\ 0 & \omega = 0 \\ j & \omega < 0 \end{cases} \tag{5.71}$$

This can be interpreted as an **all-pass** filter, providing a phase shift of $\pi/2$ radians at all frequencies. Hence, if the input signal is $\cos(\omega t)$, the output will be $\cos(\omega t - 90) = \sin(\omega t)$.

The corresponding impulse response is

$$h(t) = \frac{1}{\pi t} \quad \text{for } t \neq 0 \tag{5.72}$$

The response of the quadrature filter to a real input signal is denoted the **Hilbert Transform** and can be expressed as the convolution of the input signal $x(t)$ and the impulse response of the filter

$$y(t) = x(t) * h(t) = \frac{1}{\pi} \int_{-\infty}^{\infty} \frac{x(\tau)}{t - \tau} d\tau \tag{5.73}$$

where * denotes convolution. Now, the complex **analytic signal** $z(t)$ associated with $x(t)$ can be formed

$$z(t) = x(t) + jy(t) = x(t) + jx(t) * h(t) \tag{5.74}$$

It is clear that $z(t)$ above is the response of the system

$$G(\omega) = 1 + jH(\omega) \tag{5.75}$$

The frequency response above implies attenuation of 'negative' frequency components while passing 'positive' frequency components. In a SSB situation, this relates to the lower and upper sideband (Proakis, 1989) respectively

$$G(\omega) = \begin{cases} 1 + j(-j) = 2 & \text{for } \omega > 0 \\ 1 + j(j) = 0 & \text{for } \omega < 0 \end{cases} \tag{5.76}$$

Hence, if we want to transmit the analog signal $x(t)$ using SSB modulation equation (5.74) replaces equation (5.67). Further, the symbol mapper in Figure 5.5 is replaced by a direct connection of $x(t)$ to the I channel and a connection via a quadrature filter (Hilbert transformer) to the Q channel (the phase shifter connected to the carrier generator in Figure 5.5 can of course also be implemented as a Hilbert transformer.)

When dealing with discrete time systems, the impulse response (5.72) is replaced by

$$h(n) = \begin{cases} 1/\pi n & \text{for odd } n \\ 0 & \text{for even } n \end{cases} \tag{5.77}$$

This function can be approximated by for instance a FIR filter. There are different ways of approximating the time-discrete Hilbert transformer (Mitra and Kaiser, 1993).

6 Introduction to Kalman filters

6.1 An intuitive approach

Filters were originally viewed as systems or algorithms with frequency selective properties. They could discriminate between unwanted and desired signals found in different frequency bands. Hence, by using a filter, these unwanted signals could be rejected.

In the 1940s, theoretical work was performed by N. Wiener and N.A. Kolomogorov on statistical approaches to filtering. In these cases signals could be filtered according to their **statistical properties**, rather than their frequency contents. At this time the Wiener and Kolomogorov theory required the signals to be **stationary**, which means that the statistical properties of the signals were not allowed to change with time.

In late 1950s and early 1960s a new theory was developed capable of coping with **non-stationary** signals as well. This theory came to be known as the **Kalman filter theory**, named after R.E. Kalman.

The Kalman filter is a **discrete time**, **linear filter** and it is also an **optimal filter** under conditions that shall be described later in the text. The underlying theory is quite advanced and is based on statistical results and estimation theory. This presentation of the Kalman filter (Åström and Wittenmark, 1984; Anderson and Moore, 1979) will begin with a brief discussion of estimation.

6.1.1 Recursive least-square estimation

Let us start with a simple example to illustrate the idea of the **recursive least-square (RLS) estimation**. Assume that we would like to measure a constant signal level (DC level) x. Unfortunately, our measurements $z(n)$ are disturbed by noise $v(n)$. Hence, based on the observations

$$z(n) = x + v(n) \tag{6.1}$$

Our task is to filter out the noise and make the best possible estimation of x. This estimate is called $\hat{x}$. Our quality criterion is finding the estimate $\hat{x}$ that minimizes the least-square criteria

$$J(\hat{x}) = \sum_{n=1}^{N} \left(z(n) - \hat{x} \right)^2 \tag{6.2}$$

The minimum of equation (6.2) can be found by setting the first derivative to zero

$$\frac{\partial J(\hat{x})}{\partial \hat{x}} = \sum_{n=1}^{N} 2(\hat{x} - z(n)) = 0 \tag{6.3}$$

Solving for $\hat{x}(N)$, that is, the estimate $\hat{x}$ used during N observations of $z(n)$ we obtain

$$\sum_{n=1}^{N} \hat{x} = \sum_{n=1}^{N} z(n) \tag{6.4}$$

where

$$\sum_{n=1}^{N} \hat{x} = N\hat{x}(N) = \sum_{n=1}^{N} z(n) \tag{6.5}$$

Hence, the best way to filter our observations in the sense of minimizing our criteria (6.2) is

$$\hat{x}(N) = \frac{1}{N} \sum_{n=1}^{N} z(n) \tag{6.6}$$

that is, taking the average of N observations. Now, we do not want to wait for N observations before obtaining an estimate. We want to have an estimate almost at once, when starting to measure. Of course, this estimate will be quite poor, but we expect it to grow better and better, as more observations are gathered. In other words, we require a **recursive** estimation (filtering) method. This can be achieved in the following way:

From equation (6.5) we can express the estimate after $N+1$ observations

$$(N+1)\hat{x}(N+1) = \sum_{n=1}^{N+1} \hat{x} = \sum_{n=1}^{N+1} z(n) \tag{6.7}$$

Now, we want to know how the estimate changes from N to $N + 1$ observations, in order to find a recursive algorithm. Taking equation (6.7) minus equation (6.5) gives

$$(N+1)\hat{x}(N+1) - N\hat{x}N = \sum_{n=1}^{N+1} z(n) - \sum_{n=1}^{N} z(n) = z(N+1) \tag{6.8}$$

Rearranging in a way that $\hat{x}(N+1)$ (the 'new' estimate) is expressed in terms of $\hat{x}(N)$ (the 'old' estimate), from equation (6.8) we get

$$\hat{x}(N+1) = \frac{1}{N+1}\left(z(N+1) + N\hat{x}(N)\right)$$

$$= \frac{1}{N+1}\left(z(N+1) + (N+1)\hat{x}(N) - \hat{x}(N)\right)$$

$$= \hat{x}(N) + \frac{1}{N+1}\left(z(N+1) - \hat{x}(N)\right) \tag{6.9}$$

where we assume the initial condition $\hat{x}(0) = 0$. The filter can be drawn as in Figure 6.1. Note that the filter now contains a **model** (the delay z^{-1} or

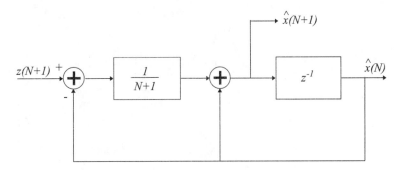

Figure 6.1 *An example of a recursive least-square estimator*

'memory') of the signal x. Presently, the model now holds the best estimate $\hat{x}(N)$ which is compared to the new measured value $z(N+1)$. The difference $z(N+1) - \hat{x}(N)$, sometimes called **the innovation**, is amplified by the **gain factor** $1/(N+1)$ before it is used to update our estimate, that is our model, to $\hat{x}(N+1)$.

Two facts need to be noted. Firstly, if our estimate $\hat{x}(N)$ is close to the measured quantity $z(N+1)$ our model is not adjusted significantly and the output is quite good as is. Secondly, the gain factor $1/(N+1)$ will grow smaller and smaller as time goes by. The filter will hence pay less and less attention to the measured noisy values, making the output level stabilize to a fixed value of $\hat{x}$.

In the discussion above, we have assumed that the impact of the noise has been constant. The 'quality' of the observed variable $z(n)$ has been the same for all n. If, on the other hand we know that the noise is very strong at certain times, or even experience a disruption in the measuring process, then we should of course pay less attention to the input signal $z(n)$. One way to solve this is to implement a kind of quality weight $q(n)$. For instance, this weight could be related to the magnitude of the noise as

$$q(n) \propto \frac{1}{v^2(n)} \tag{6.10}$$

Inserting this into equation (6.2) we obtain a **weighted least-square** criteria

$$J(\hat{x}) = \sum_{n=1}^{N} q(n)\, (z(n) - \hat{x})^2 \tag{6.11}$$

Using equation (6.11) and going through the same calculations as before (equations (6.3) to (6.9)) we obtain the expression for the gain factor in this case (compare to equation (6.9))

$$\hat{x}(N+1) = \hat{x}(N) + \frac{q(N+1)}{\sum_{n=1}^{N+1} q(n)} \left(z(N+1) - \hat{x}(N) \right)$$

$$= \hat{x}(N) + Q(N+1)\left(z(N+1) - \hat{x}(N) \right) \tag{6.12}$$

It can be shown that the gain factor $Q(N+1)$ can also be calculated recursively

$$Q(N+1) = \frac{q(N+1)}{\sum_{n=1}^{N+1} q(n)} = \frac{q(N+1)}{\sum_{n=1}^{N} q(n) + q(N+1)}$$

$$= \frac{q(N+1)}{\frac{q(N)}{Q(N)} + q(N+1)} = \frac{Q(N)q(N+1)}{q(N) + Q(N)q(N+1)} \tag{6.13}$$

where the starting conditions are: $q(0) = 0$ and $Q(0) = 1$. We can now draw some interesting conclusions about the behaviour of the filter. If the input signal quality is extremely low (or if no input signal is available) $q(n) \to 0$ implies that $Q(n+1) \to 0$ and the output from the filter equation (6.12) is

$$\hat{x}(N+1) = \hat{x}(N) \tag{6.14}$$

In other words, we are running 'dead reckoning' using the model in the filter only. If on the other hand, the input signal quality is excellent (no noise present), $q(n) \to \infty$ and $Q(n+1) \to 1$, then

$$\hat{x}(N+1) = z(N+1) \tag{6.15}$$

In this case, the input signal is fed directly to the output and the model in the filter is updated.

For intermediate quality levels of the input signal, a mix of measured signal and modelled signal is presented at the output of the filter. This mix thus represents the best estimate of x, according to the weighted least-square criteria.

So far, we have assumed x to be a **constant**. If $x(n)$ is allowed to change over time, but is considerably slower than the noise, we must introduce a 'forgetting factor' into equation (6.2) or else all 'old' values of $x(n)$ will counteract changes of $\hat{x}(N)$. The forgetting factor $w(n)$ should have the following property

$$w(n) > w(n-1) > \cdots w(1) \tag{6.16}$$

In other words, the 'oldest' values should be forgotten the most. One example is

$$w(n) = \alpha^{N-n} \tag{6.17}$$

where $0 < \alpha < 1$. Inserting equation (6.17) into equation (6.2) and going through the calculations (6.3)–(6.9) again, we obtain

$$\hat{x}(N+1) = \hat{x}(N) + \frac{1}{\sum_{n=1}^{N+1} \alpha^{N+1-n}} \left(z(N+1) - \hat{x}(N) \right)$$

$$= \hat{x}(N) + P(N+1)\left(z(N+1) - \hat{x}(N) \right) \tag{6.18}$$

The gain factor $P(n)$ can be calculated recursively by

$$P(N+1) = \frac{P(N)}{\alpha + P(N)}$$ (6.19)

Note, the gain factors can be calculated off line, in advance.

6.1.2 The Pseudoinverse

If we now generalize the scalar measurement problem outlined above and go into a multi-dimensional problem, we can reformulate equation (6.1) using vector notation

$$z(n) = H^T(n)x + v(n)$$ (6.20)

where the dimensions of the entities are

$z(n)$: $(K \times 1)$ $H(n)$: $(L \times K)$
x: $(L \times 1)$ $v(n)$: $(K \times 1)$

In a simple case, where there is no noise present, that is $v(n) = 0$ (or when the noise is known), solving x from equation (6.20) could be done by simply inverting the matrix $H(n)$

$$x = (H^T(n))^{-1}(z(n) - v(n))$$ (6.21)

This corresponds to solving a system of L equations. Inverting the matrix $H(n)$ may not be possible in all cases for two reasons, either the inverse does not exist or we do not have access to a sufficient amount of information. The latter is the case after each observation of $z(n)$ when noise $v(n)$ is present or if $K < L$, that is we have too few observations. Hence, a straight-forward matrix inversion is not possible. In such a case, where the inverse of the matrix cannot be found, a **pseudoinverse** (Åström and Wittenmark, 1984; Anderson and Moore, 1979) also called the Moore–Penrose-inverse, may be used instead. Using the pseudoinverse equation (6.21) can be solved 'approximately'.

For a true inverse, $(H^T)^{-1}H^T = I$, where I is the identity matrix. In a similar way, using the pseudoinverse $H^\#$ we try to make the matrix product $(H^T)^\# H^T$ as close to the identity matrix as possible in a least-squares sense. Note, this is similar to fitting a straight line to a number of measured and plotted points in a diagram using the least-square method.

The pseudoinverse of $H(n)$ can be expressed as

$$H^\# = (H^T H)^{-1} H^T$$ (6.22)

Now, finding the pseudoinverse corresponds to minimizing the criteria (compare to equation (6.2))

$$J(\hat{x}) = \sum_{n=1}^{N} \| z(n) - H^T(n)\hat{x} \|^2$$ (6.23)

where the Euclidean matrix norm is used, that is

$$\|A\|^2 = \sum_i \sum_j a_{ij}^2$$

Minimizing the above is finding the vector estimate $\hat{x}$ that results in the least-square error. Taking the derivative of equation (6.23) for all components, we obtain the gradient. Similar to equations (6.3)–(6.5), we solve for $\hat{x}$ as the gradient is set equal to zero, thus obtaining the following (compare to equation (6.6))

$$\hat{x}(N) = \left(\sum_{n=1}^{N} H(n)H^T(n) \right)^{-1} \sum_{n=1}^{N} H(n)z(n) \tag{6.24}$$

Note that the sums above constitute the pseudoinverse. In the same way as before, we like to find a recursive expression for the best estimate. This can be achieved by going through calculations similar to equations (6.7) and (6.8), but using vector and matrix algebra. These calculations will yield the result (compare to equation (6.9))

$$\hat{x}(N+1) = \hat{x}(N) + \left(\sum_{n=1}^{N+1} H(n)H^T(n) \right)^{-1} H(N+1)\left(z(N+1) - H^T(N+1)\hat{x}(N)\right)$$

$$= \hat{x}(N) + P(N+1)H(N+1)\left(z(N+1) - H^T(N+1)\hat{x}(N)\right) \tag{6.25}$$

where $P(n)$ is the gain factor as before. Reasoning in a similar way to the calculations of equation (6.13), we find that this gain factor can also be calculated recursively

$$P(N+1) = P(N) - P(N)H(N+1)$$

$$\left(I + H^T(N+1)P(N)H(N+1)\right)^{-1}H^T(N+1)P(N) \tag{6.26}$$

The equations (6.25) and (6.26) comprise a recursive method of obtaining the pseudoinverse $H^{\#}$ using a filter model as in Figure 6.1. The ideas presented above constitute the underlying ideas of Kalman filters.

6.2 The Kalman filter

6.2.1 The signal model

The **signal model**, sometimes also called the **process model** or the **plant**, is a model of the 'reality' which we would like to measure. This 'reality' also generates the signals we are observing. In this context, dynamic systems are commonly described using a state–space model (see Chapter 1). A simple example may be the following.

Assume that our friend Bill is cruising down Main Street in his brand new Corvette. Main Street can be approximated by a straight line and since Bill has engaged the cruise control, we can assume that he is travelling at a constant speed (no traffic lights). Using a simple radar device, we try to measure Bill's position along Main Street (starting from 'Burger King') at every second.

Now, let us formulate a discrete time, **state–space model**. Let Bill's position at time n be represented by the discrete time variable $x_1(n)$ and his speed by $x_2(n)$. Expressing this in terms of a recursive scheme we can write

$$x_1(n+1) = x_1(n) + x_2(n)$$

$$x_2(n+1) = x_2(n)$$

(6.27)

The second equation simply tells us that the speed is constant. The equations above can also be written using vector notation by defining a **state vector** $x(n)$ and a **transition matrix** $F(n)$, as

$$x(n) = \begin{bmatrix} x_1(n) \\ x_2(n) \end{bmatrix} \quad \text{and} \quad F(n) = \begin{bmatrix} 1 & 1 \\ 0 & 1 \end{bmatrix}$$

(6.28)

The equations (6.27) representing the system can now be nicely formulated as a simple state–space model

$$x(n+1) = F(n)x(n)$$

(6.29)

This is of course quite a trivial situation and an ideal model, but Bill is certainly not. Now and then, he brakes a little, when there is something interesting at the sidewalk. This repeated braking and putting the cruise control back into gear changes the speed of the car. If we assume that the braking positions are randomly distributed along Main Street, we can hence take this into account by adding a white Gaussian noise signal $w(n)$ to the speed variable in our model

$$x(n+1) = F(n)x(n) + G(n)w(n)$$

(6.30)

where

$$G(n) = \begin{bmatrix} 0 \\ 1 \end{bmatrix}$$

The noise, sometimes called 'process noise' is supposed to be scalar in this example, having the variance: $Q = \sigma_w^2$ and a mean equal to zero. Note, $x(n)$ is now a stochastic vector.

So far, we have not considered the errors of the measuring equipment. What entities in the process (Bill and Corvette) can we observe using our simple radar equipment? To start with, since the equipment is old and not of Doppler type, it will only give a number representing the **distance**. Speed is not measured. Hence, we can only get information about the state variable $x_1(n)$. This is represented by the observation matrix $H(n)$. Further, there are of course random errors present in the distance measurements obtained. On some occasions, no sensible readings are obtained, when cars are crossing the street. This uncertainty can be modelled by adding another white Gaussian noise signal $v(n)$. This so-called 'measurement noise' is scalar in this example, and is assumed to have zero mean and a variance $R = \sigma_v^2$. Hence, the measured signal $z(n)$ can be expressed as

$$z(n) = H^T(n)x(n) + v(n)$$

(6.31)

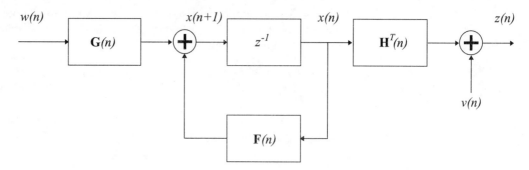

Figure 6.2 *Basic signal model as expressed by equations (6.30) and (6.31)*

Equations (6.30) and (6.31) now constitute our basic signal model, which can be drawn as in Figure 6.2.

6.2.2 The filter

The task of the filter, given the observed signal $z(n)$ (a vector in the general case), is to find the best possible estimate of the state vector $x(n)$ in the sense of the criteria given below. We should however remember that, $x(n)$ is now a stochastic signal rather than a constant, as in the previous section.

For our convenience, we will introduce the following notation: the estimate of $x(n)$ at time n, based on the $n-1$ observations $z(0), z(1), \ldots z(n-1)$, will be denoted $\hat{x}(n|n-1)$ and the set of observations $z(0), z(1), \ldots z(n-1)$ itself will be denoted $Z(n-1)$.

Our quality criterion is finding the estimate that minimizes the conditional error covariance matrix (Anderson and Moore, 1979)

$$C(n|n-1) = \mathrm{E}\big[(x(n) - \hat{x}(n|n-1))(x(n) - \hat{x}(n|n-1))^T | Z(n-1)\big] \quad (6.32)$$

This is a **minimum variance** criterion and can be regarded as a kind of 'stochastic version' of the least-square criteria used in the previous section. Finding the minimum is a bit more complicated in this case than in the previous section. The best estimate, according to our criteria, is found using the **conditional mean** (Anderson and Moore, 1979), that is

$$\hat{x}(n|n) = \mathrm{E}\big[x(n)|Z(n)\big] \quad (6.33)$$

The underlying idea is as follows: $x(n)$ and $z(n)$ are both random vector variables of which $x(n)$ can be viewed as being a 'part' of $z(n)$ (see equation (6.31)). The statistical properties of $x(n)$ will be 'buried' in the statistical properties of $z(n)$. For instance, if we now want to have a better estimate of the mean of $x(n)$, uncertainty can be reduced by considering the actual values of the measurements $Z(n)$. This is called **conditioning**. Hence equation (6.33), the conditional mean of $x(n)$, is the **most probable** mean of $x(n)$, given the measured values $Z(n)$.

We will show how this conditional mean (6.33) can be calculated recursively, which is exactly what the **Kalman filter** does (later we return to the example of Bill and his Corvette).

Since we are going for a recursive procedure, let us start at time $n = 0$, when there are no measurements made. We have (equation (6.32))

$$C(0|-1) = E[x(0)x^T(0)|Z(-1)] = P(0) \tag{6.34}$$

It can be shown (Anderson and Moore, 1979) that the conditional mean of $x(0)$ can be obtained from the cross-covariance of $x(0)$ and $z(0)$, the auto-covariance and the mean of $z(0)$ and the measured value $z(0)$ itself

$$\hat{x}(0|0) = E[x(0)] + C_{xz}(0)C_{zz}^{-1}(0)(z(0) - E[z(0)]) \tag{6.35}$$

The covariance matrix at time n, based on n observations is denoted $C(n|n)$. It can be obtained (for $n = 0$) from

$$C(0|0) = C_{xx}(0) - C_{xz}(0)C_{zz}^{-1}(0)C_{zx}(0) \tag{6.36}$$

Next, we need to find the mean vectors and covariance matrices to plug into equations (6.35) and (6.36). Both the process noise and the measurement noise are Gaussian, and we assume that they are uncorrelated.

The mean of $x(0)$ is denoted $E[x(0)]$. Using equation (6.31), the mean of $z(0)$ is

$$E[z(0)] = E[H^T(0)x(0) + v(0)] = H^T(0)E[x(0)] \tag{6.37}$$

where we have used the fact that the mean of the measurement noise $v(0)$ is zero.

The auto-covariance of $x(0)$ is

$$C_{xx}(0) = E[x(0)x^T(0)] = P(0) \tag{6.38}$$

Using equation (6.31), the auto-covariance of $z(0)$ can be expressed as

$$\begin{aligned}
C_{zz}(0) &= E[z(0)z^T(0)] \\
&= E[(H^T(0)x(0) + v(0))(H^T(0)x(0) + v(0))^T] \\
&= E[H^T(0)x(0)x^T(0)H(0) + H^T(0)x(0)v^T(0) \\
&\quad + v(0)x^T(0)H(0) + v(0)v^T(0)] \\
&= H^T(0)E[x(0)x^T(0)]H(0) + E[v(0)v^T(0)] \\
&= H^T(0)P(0)H(0) + R(0) \tag{6.39}
\end{aligned}$$

where the cross-covariance between $x(0)$ and $v(0)$ is zero, since the measurement noise was assumed to be uncorrelated to the process noise. $R(0)$ is the auto-correlation matrix of the measurement noise $v(0)$. In a similar way, the cross-covariance between $x(0)$ and $z(0)$ can be expressed as

$$C_{zx}(0) = \mathrm{E}\big[z(0)x^T(0)\big]$$

$$= \mathrm{E}\big[(H^T(0)x(0) + v(0))x^T(0)\big]$$

$$= \mathrm{E}\big[H^T(0)x(0)x^T(0) + v(0)x^T(0)\big]$$

$$= H^T(0)\,\mathrm{E}\big[x(0)x^T(0)\big] = H^T(0)P(0) \tag{6.40}$$

and

$$C_{xz}(0) = \mathrm{E}\big[x(0)z^T(0)\big] = P(0)H(0) \tag{6.41}$$

Inserting these results into equations (6.35) and (6.36) respectively, we obtain the conditioned mean and the covariance

$$\hat{x}(0|0) = \mathrm{E}\big[x(0)\big] + P(0)H(0)\big(H^T(0)P(0)H(0) + R(0)\big)^{-1}$$
$$\big(z(0) - H^T(0)\mathrm{E}\big[x(0)\big]\big) \tag{6.42}$$

$$C(0|0) = P(0) - P(0)H(0)\big(H^T(0)P(0)H(0) + R(0)\big)^{-1}H^T(0)P(0) \tag{6.43}$$

Let us now take a step forward in time, that is for $n = 1$, before we have taken the measurement $z(1)$ into account, from equation (6.30) and various independence assumptions (Anderson and Moore, 1979) we have the mean

$$\hat{x}(1|0) = F(0)\hat{x}(0|0) \tag{6.44}$$

and the covariance

$$C(1|0) = F(0)C(0|0)F^T(0) + G(0)Q(0)G^T(0) \tag{6.45}$$

where $Q(0)$ is the auto-correlation matrix of the process noise $w(0)$. We are now back in the situation where we started for $n = 0$, but now with $n = 1$. The calculations starting with equation (6.35) can be repeated to take the measurement $z(1)$ into account, that is to do the conditioning. For convenience, only the last two equations in the sequel will be shown (compare to equations (6.42) and (6.43)).

$$\hat{x}(1|1) = \hat{x}(1|0) + C(1|0)H(1)\big(H^T(1)C(1|0)H(1) + R(1)\big)^{-1}$$
$$\big(z(1) - H^T(1)\hat{x}(1|0)\big) \tag{6.46}$$

$$C(1|1) = C(1|0) - C(1|0)H(1)\big(H^T(1)C(1|0)H(1) + R(1)\big)^{-1}$$
$$H^T(1)C(1|0) \tag{6.47}$$

Repeating the steps as outlined above, we can now formulate a general, recursive algorithm that constitutes the **Kalman filter equations** in their basic form. These equations are commonly divided into two groups, the **measurement-update equations** and the **time-update equations**.

Measurement-update equations

$$\hat{x}(n|n) = \hat{x}(n|n-1) + C(n|n-1)H(n)\big(H^T(n)C(n|n-1)H(n) + R(n)\big)^{-1}$$

$$\big(z(n) - H^T(n)\hat{x}(n|n-1)\big) \tag{6.48}$$

$$C(n|n) = C(n|n-1) - C(n|n-1)H(n)$$

$$\big(H^T(n)C(n|n-1)H(n) + R(n)\big)^{-1}H^T(n)C(n|n-1) \tag{6.49}$$

Time-update equations

$$\hat{x}(n+1|n) = F(n)\hat{x}(n|n) \tag{6.50}$$

$$C(n+1|n) = F(n)C(n|n)F^T(n) + G(n)Q(n)G^T(n) \tag{6.51}$$

The equations above are straightforward to implement in software. An alternative way of writing them, making it easier to draw the filter in diagram form is

$$\hat{x}(n+1|n) = F(n)\hat{x}(n|n-1) + K(n)\big(z(n) - H^T(n)\hat{x}(n|n-1)\big) \tag{6.52}$$

where $K(n)$ is the **Kalman gain** matrix

$$K(n) = F(n)C(n|n-1)H(n)\big(H^T(n)C(n|n-1)H(n) + R(n)\big)^{-1} \tag{6.53}$$

and the conditional error covariance matrix is given recursively by a discrete time **Riccati equation**

$$C(n+1|n) = F(n)\big(C(n|n-1) - C(n|n-1)H(n)$$

$$\big(H^T(n)C(n|n-1)H(n) + R(n)\big)^{-1}H^T(n)C(n|n-1)\big)F^T(n)$$

$$+ G(n)Q(n)G^T(n) \tag{6.54}$$

The structure of the corresponding Kalman Filter is shown in Figure 6.3. The filter can be regarded as a copy of the signal model (see Figure 6.2) put in a feedback loop. The input to the model is the difference between the actual measured signal $z(n)$ and our estimate of the measured signal $H^T(n)\hat{x}(n|n-1)$ multiplied by the Kalman gain $K(n)$. This is largely the same approach as the recursive least-square estimator in Figure 6.1. The difference

$$\tilde{z}(n) = z(n) - H^T(n)\hat{x}(n|n-1) \tag{6.55}$$

is sometimes referred to as the **innovation**.

Now, to conclude this somewhat simplified presentation of the Kalman filter, let us go back to Bill and the Corvette. As an example, we will design a Kalman filter to estimate Bill's speed and position from the not-too-good measurements available from the radar equipment.

We have earlier defined the signal model, that is, we know $x(n)$, F and G. We are now writing the matrices without indices, since they are constants. Further, we recall that the process noise is a Gaussian random variable $w(n)$,

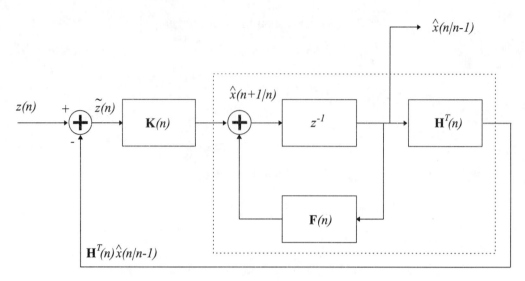

Figure 6.3 *A basic Kalman filter. Note the copy of the signal model in the dotted area*

which in this example is a scalar resulting in the auto-correlation matrix Q turning into a scalar as well. This is simply the variance $Q = \sigma_w^2$ which is assumed to be constant over time.

Regarding the measurement procedure, since we can only measure position and not speed, the observation matrix H (constant) will be (see also equation (6.31))

$$H = \begin{bmatrix} 1 \\ 0 \end{bmatrix} \tag{6.56}$$

Since the measured signal $z(n)$ is a scalar in this example, the measurement noise $v(n)$ will also be a scalar, and the auto-correlation matrix R will simply consist of the constant variance $R = \sigma_v^2$. Now, all the details regarding the signal model are determined, but we also need some vectors and matrices to implement the Kalman filter.

First, we need vectors for the estimates of $x(n)$

$$\hat{x}(n|n-1) = \begin{bmatrix} \hat{x}_1(n|n-1) \\ \hat{x}_2(n|n-1) \end{bmatrix}$$

Then, we will need matrices for the error covariance

$$C(n|n-1) = \begin{bmatrix} C_{11}(n|n-1) & C_{12}(n|n-1) \\ C_{21}(n|n-1) & C_{22}(n|n-1) \end{bmatrix}$$

If we insert all the known facts above into equations (6.48)–(6.51), the Kalman filter for this example is defined. From equation (6.48), the measurement-update yields

$$\hat{x}(n|n) = \hat{x}(n|n-1) + \begin{bmatrix} C_{11}(n|n-1) \\ C_{21}(n|n-1) \end{bmatrix} \frac{(z(n) - \hat{x}_1(n|n-1))}{(C_{11}(n|n-1) + R)} \tag{6.57}$$

which can be expressed in component form

$$\hat{x}_1(n|n) = \hat{x}_1(n|n-1) + \frac{C_{11}(n|n-1)(z(n) - \hat{x}_1(n|n-1))}{(C_{11}(n|n-1) + R)} \tag{6.58a}$$

$$\hat{x}_2(n|n) = \hat{x}_2(n|n-1) + \frac{C_{21}(n|n-1)(z(n) - \hat{x}_1(n|n-1))}{(C_{11}(n|n-1) + R)} \tag{6.58b}$$

Using equation (6.49) the covariance can be updated

$$C(n|n) = C(n|n-1)$$

$$- \begin{bmatrix} C_{11}(n|n-1)C_{11}(n|n-1) & C_{11}(n|n-1)C_{12}(n|n-1) \\ C_{21}(n|n-1)C_{11}(n|n-1) & C_{21}(n|n-1)C_{12}(n|n-1) \end{bmatrix}$$

$$\frac{1}{(C_{11}(n|n-1) + R)} \tag{6.59}$$

or expressed in component form

$$C_{11}(n|n) = C_{11}(n|n-1) - \frac{(C_{11}(n|n-1))^2}{(C_{11}(n|n-1) + R)} \tag{6.60a}$$

$$C_{12}(n|n) = C_{12}(n|n-1) - \frac{C_{11}(n|n-1)C_{12}(n|n-1)}{(C_{11}(n|n-1) + R)} \tag{6.60b}$$

$$C_{21}(n|n) = C_{21}(n|n-1) - \frac{C_{21}(n|n-1)C_{11}(n|n-1)}{(C_{11}(n|n-1) + R)} \tag{6.60c}$$

$$C_{22}(n|n) = C_{22}(n|n-1) - \frac{C_{21}(n|n-1)C_{12}(n|n-1)}{(C_{11}(n|n-1) + R)} \tag{6.60d}$$

Then, the time-update equations, starting with equation (6.50)

$$\hat{x}(n+1|n) = \begin{bmatrix} \hat{x}_1(n|n) + \hat{x}_2(n|n) \\ \hat{x}_2(n|n) \end{bmatrix} \tag{6.61}$$

This can be compared to the basic state–space signal model (6.27). If equation (6.61) is expressed in component form, we get

$$\hat{x}_1(n+1|n) = \hat{x}_1(n|n) + \hat{x}_2(n|n) \tag{6.62a}$$

$$\hat{x}_2(n+1|n) = \hat{x}_2(n|n) \tag{6.62b}$$

Finally, equation (6.51) gives

$$C(n+1|n) =$$

$$\begin{bmatrix} C_{11}(n|n) + C_{21}(n|n) + C_{12}(n|n) + C_{22}(n|n) & C_{12}(n|n) + C_{22}(n|n) \\ C_{21}(n|n) + C_{22}(n|n) & C_{22}(n|n) \end{bmatrix}$$

$$+ \begin{bmatrix} 0 & 0 \\ 0 & Q \end{bmatrix} \tag{6.63}$$

and in component form

$$C_{11}(n+1|n) = C_{11}(n|n) + C_{21}(n|n) + C_{12}(n|n) + C_{22}(n|n) \tag{6.64a}$$

$$C_{12}(n+1|n) = C_{12}(n|n) + C_{22}(n|n) \tag{6.64b}$$

$$C_{21}(n+1|n) = C_{21}(n|n) + C_{22}(n|n) \tag{6.64c}$$

$$C_{22}(n+1|n) = C_{22}(n|n) + Q \tag{6.64d}$$

Hence, the Kalman filter for estimating Bill's position and speed is readily implemented by calculating repetitively the 12 equations and (6.58a), (6.58b), (6.60a)–(6.60d), (6.62a) and (6.62b) and (6.64a)–(6.64d) above.

For this example, Figure 6.4 shows the true velocity and position of the car and the noisy, measured position, that is the input to the Kalman filter. Figure 6.5 shows the output from the filter, the estimated velocity and position. An overshot can be seen in the beginning of the filtering process, before the filter is tracking. Figure 6.6 shows the two components of the decreasing Kalman gain as a function of time.

6.2.3 Kalman filter properties

At first it should be stressed that the brief presentation of the Kalman filter in the previous section is simplified. For instance, the assumption about Gaussian noise is not necessary in the general case (Anderson and Moore, 1979). Nor is the assumption that the process and measurement noise is uncorrelated. There are also a number of extensions (Anderson and Moore, 1979) of the Kalman filter which have not been described here. Below, we will however discuss some interesting properties of the general Kalman filter.

The Kalman filter is linear. This is obvious from the preceding calculations. The filter is also a **discrete-time** system and has **finite dimensionality**.

The Kalman filter is an optimal filter in the sense of achieving minimum variance estimates. It can be shown that in Gaussian noise situations, the Kalman filter is the best possible filter, and in non-Gaussian cases, the **best** linear (Anderson and Moore, 1979) filter. The optimal filter in the latter case is non-linear and may therefore be very hard to find and analyse in the general case.

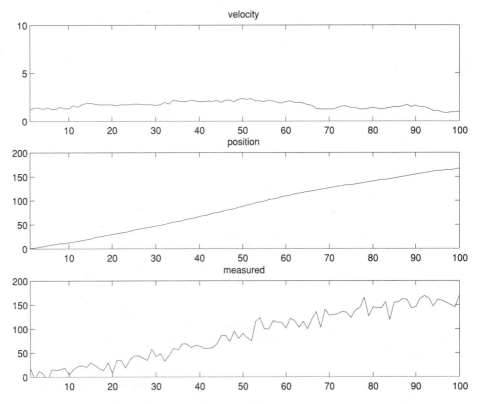

Figure 6.4 *True velocity and position and noisy measured position*

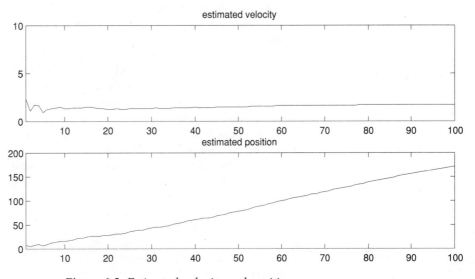

Figure 6.5 *Estimated velocity and position*

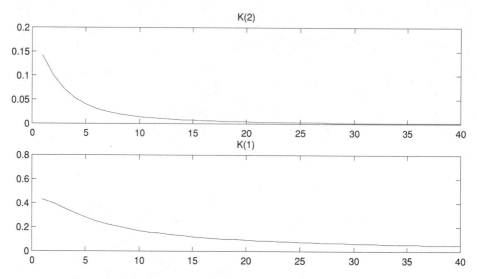

Figure 6.6 *Kalman gain as a function of time*

**The gain matrix $K(n)$ can be calculated off line, in advance, before
the Kalman filter is actually run**. The output $\hat{x}(n|n-1)$ of the filter obvi-
ously depends on the input $z(n)$, but the covariance matrix $C(n|n-1)$ and
hence the Kalman gain $K(n)$ does **not**. $K(n)$ can be regarded as the smartest
way of taking the measurements into account given the signal model and
statistical properties.

From above we can also conclude that since $C(n|n-1)$ is independent of
the measurements $z(n)$, no one set of measurements helps more than any
other to eliminate the uncertainty about $x(n)$.

Another conclusion that can be drawn from above is that the filter is only
optimal given the signal model and statistical assumptions made at design
time. If there is a poor match between the real world signals and the assumed
signals of the model, the filter will of course not perform optimally in reality.
This problem is however common to all filters.

Further, even if the signal model is time invariant and the noise processes
are stationary, that is $F(n)$, $G(n)$, $H(n)$ $Q(n)$ and $R(n)$ are constant, in general
$C(n|n-1)$ and hence $K(n)$ will **not** be constant. This implies that in the
general case, the **Kalman filter will be time varying**.

The Kalman filter contains a **model**, which **tracks** the true system we are
observing. So, from the model, we can obtain estimates of state variables
that we are only measuring in an indirect way. We could for instance, get
an estimate of the speed of Bill's Corvette in our example above, despite
only measuring the position of the car.

Another useful property of the built-in model is that in the case of missing
measurements during a limited period of time, the filter can '**interpolate**'
the state variables. In some applications, when the filter model has 'stabil-
ized', it can be 'sped up' and even be used for **prediction**.

6.2.4 Applications

The Kalman filter is a very useful device that has found many applications in diverse areas. Since the filter is a discrete-time system, the advent of powerful and not-too-costly DSP circuits has been crucial to the possibilities of using Kalman filters in commercial applications.

The Kalman filter is used for **filtering** and **smoothing** of measured signals, not only in electronic applications, but also in processing of data in the areas of economy, medicine, chemistry and sociology, for example. Kalman filters are also, to some extent, used in **digital image processing**, when enhancing the quality of digitized pictures. Further detection of signals in radar and sonar systems and in telecommunication systems often require filters for equalization. The Kalman filter, belonging to the class of adaptive filters performs well in such contexts (see Chapter 3).

Other areas where Kalman filters are used are **process identification**, **modelling** and **control**. Much of the early developments of the Kalman filter theory came from applications in the aerospace industry. One control system example is keeping satellites or missiles on a desired trajectory. This task is often solved using some optimal control algorithm, taking estimated state variables as inputs. The estimation is of course done by a Kalman filter. The estimated state vector may be of dimension 12 (or more) consisting of position (x, y, z), yaw, roll, pitch and the first derivatives of these, i.e. speed (x, y, z) and speed of yaw, roll and pitch movements. Designing such a system requires a considerable amount of computer simulations.

An example of process modelling using Kalman filters is to analyse the behaviour of the stock market, and/or to find parameters for a model of the underlying economic processes. Modelling of meteorological and hydrological processes as well as chemical reactions in the manufacturing industry are other examples.

Kalman filters can also be used for **forecasting**, for instance prediction of air pollution levels, air traffic congestion etc.

7 Data compression

7.1 An information theory primer

7.1.1 Historic notes

The concept of **information theory** (Cover and Thomas, 1991) in a strict mathematical sense was born in 1948, when C.E. Shannon published his celebrated work *A Mathematical Theory of Communication*, later published using the title *The Mathematical Theory of Communication*. Obviously, Shannon was ahead of his time and many of his contemporary communication specialists did not understand his results. Gradually however it became apparent that Shannon had indeed created a new scientific discipline.

Besides finding a way of quantizing information in a mathematical sense, Shannon formulated three important, fundamental theorems: **the source coding theorem**, **the channel coding theorem** and **the rate distortion theorem**. In this chapter, we will concentrate on the source coding theorem, while implications of the channel coding theorem will be discussed in Chapter 8.

The work of Shannon is an important foundation of modern information theory. It has proven useful not only in circuit design computer design, and communications technology, but is also being applied to biology and psychology, to phonetics and even to semantics and literature.

7.1.2 Information and entropy

There are many different definitions of the term 'information', however in this case we will define it as receiving **information** implies a **reduction of uncertainty** about a certain condition. Now, if we assume that this condition or variable can take a finite number of values or states, these states can be represented using a finite set of **symbols**, an 'alphabet'.

Consider the following example: in a group of ten persons, we need to know in which country in Scandinavia each person is born. There are three possibilities, Denmark, Norway or Sweden. We decide to use a subset of the digits 0, 1, 2, ... 9 as our symbol alphabet as follows: 1 = Denmark, 2 = Norway, 3 = Sweden. The complete **message** will be a 10 digit long string of symbols, where each digit represents the birthplace of a person. Assume that the probability of being born in any of the three countries is equal, $p_1 = 1/3$ (born in Denmark), $p_2 = 1/3$ (born in Norway) and $p_3 = 1/3$ (born in Sweden), the total uncertainty about the group (in this respect) before the message is received is one out of $3^{10} = 59049$ (the number of possible combinations). For every symbol of the message we receive, the uncertainty is reduced by a factor of 3, and when the entire message is received, there is only one possible combination left. Our uncertainty has hence been reduced and information has been received.

Extending our example above, we now need to know in what country in Europe the people are born. We realize that our symbol alphabet must be modified, since there are 34 countries in Europe today (changes rather quickly these days). We extend our alphabet in the following way. First we use 0, 1, 2, ... 9 as before, and after that we continue using the normal letters, A, B, C, ... Z. This will give us an alphabet consisting of $10 + 26 = 36$ different symbols. If all the countries are equally probable birthplaces, the uncertainty before we get the message is one out of $34^{10} \approx 2.06 \cdot 10^{15}$ possible combinations. Every symbol reduces the information by a factor of 34 and when the entire 10 symbol string is received, no uncertainty exists.

From the above example, we draw the conclusion that in the latter case, the message contained a larger amount of information than in the first case. This is because of the fact that in the latter case there were more countries to choose from than in the former. The uncertainty 'which country in Europe?' is of course larger than the uncertainty 'which country in Scandinavia?'. If we know in advance that there are only people born in Scandinavia in the group, this represents a certain amount of information that we already have and that need not be included in the message. From this discussion, we are now aware of the fact that when dealing with information measures, a very relevant question is 'information about **what**?'.

In the latter case above, when extending the alphabet, we could have chosen another approach. Instead of using **one** character per symbol, we could have used **two** characters, for example 01, 02, 03, ... 33, 34. In this way, only the digits 0 ... 9 would have been used as in the earlier case, but on the other hand, the message would have been twice as long, that is 20 characters. Actually, we could have used any set of characters to define our alphabet. There is however a tradeoff, the smaller the set of characters, the longer the message. The number of characters per symbol L needed can be expressed as

$$L = \left\lceil \frac{\log (N_S)}{\log (N_C)} \right\rceil = \left\lceil \log_{N_C} (N_S) \right\rceil \tag{7.1}$$

where N_S is the number of symbols needed, in other words the number of discrete states, N_C is the number of different characters used and $\lceil \ \rceil$ is the 'ceiling' operator (the first integer greater or equal to the argument). If we use for instance the popular **binary digits** (**BITs** for short) 0 and 1, this implies that $N_C = 2$. Using binary digits in our examples above implies

'Scandinavia': $N_S = 3 \quad L = \lceil \log_2 (3) \rceil = \lceil \text{lb} (3) \rceil = \lceil 1.58 \rceil = 2$ char/symbol

'Europe': $\quad N_S = 34 \quad L = \lceil \text{lb} (34) \rceil = \lceil 5.09 \rceil = 6$ char/symbol

From above, it can be seen that lb( ) means logarithm base 2, that is $\log_2$( ).

A binary message for the 'Scandinavia' example would be 20 bits long, while for the 'Europe' example, 60 bits. It is important to realize that the 'shorter' version (using many different characters) of the respective strings contains the same amount of information as the 'longer', binary version.

Now in an information theory context, we can see that the choice of characters or the preferred entries used to represent a certain amount of information does not matter. Using a smaller set of characters would however result in longer messages for a given amount of information. Quite often, the choice of character set is made upon the way a system is implemented. When analysing communication systems, binary characters (bits) are often assumed for convenience, even if another representation is used in the actual implementation.

Let us look into another situation; assume that we also want to know the gender of the people in the group. We choose to use symbols consisting of normal letters. The two symbols will be the abbreviations 'fe' for female and 'ma' for male. Hence, information about the entire group with respect to sex will be a 20 character long string, for example

fefemafemamamafefema.

Taking a closer look at this string of information, we realize that if every second letter is removed, we can still obtain all the information we need

ffmfmmmffm or eeaeaaaeea

would be OK. In this case, the longer string obviously contains 'unnecessary information', so-called **redundant information**, that is information that could be deducted from other parts of the message and thus not reduce our uncertainty (no 'news'). If the redundant information is removed, we can still reduce the uncertainty to a desired level and obtain the required information. In this simple example, the redundancy in the longer string is due to **dependency** between the characters. If we receive for instance the character 'f', we can be 100% sure that the next character will be 'e', so there is no uncertainty. The shorter versions of the string cannot be made any shorter, because there is no dependency between the characters. The sex of a person is independent of the sex of another person. The appearance of the characters 'f' and 'm' seems to be completely random. In this case, no redundancy is present. The example above is of course a very simple and obvious one. Redundancy is in many cases 'hidden' in more sophisticated ways.

Removing redundancy is mainly what **data compression** is all about. A very interesting question in this context is how much information is left in a message when all redundancy is removed? What maximum data compression factor is possible? This question is answered by the source coding theorem of Shannon. Unfortunately, the theorem does not tell us the smartest way of finding this maximum data compression method, it only tells us that there is a minimum message that contains exactly all the information we need.

Now, let us be a bit more formal about the above. To start with, we define the concept of **mutual information**. Assume that we want to gain information about an event A by observing the event B. Then the mutual information is defined as

$$I(A, B) = \log_b \left(\frac{P(A|B)}{P(A)} \right) \tag{7.2}$$

where we have assumed that the probability of the event A is non-zero, that is $P(A) \neq 0$ and that $P(B) \neq 0$. $P(A|B)$ is the **conditional probability**, that is, the probability of A given that B is observed. The unit will be 'bits' if logarithm base 2 (lb ( )) is used ($b = 2$) and 'nats' if the natural logarithm (ln ( )) is used ($b = e$).

Using Bayes' theorem (Papoulis, 1985) it is straightforward to show that $I(A, B)$ is symmetric with respect to A and B

$$I(A, B) \;=\; \log\left(\frac{P(A|B)}{P(A)}\right) = \log\left(\frac{P(AB)}{P(A)P(B)}\right)$$

$$=\; \log\left(\frac{P(B|A)}{P(B)}\right) = I(B, A) \tag{7.3}$$

Hence, it does not matter if B is observed to gain information about A, or vice versa. The same amount of information is obtained. This is why $I(A, B)$ is called the **mutual** information between the events A and B.

If we now assume that the events A and B are completely independent, there is no 'coupling' in between them whatsoever. In this case, $P(A|B) = P(A)$ and the knowledge of A is not increased by the fact that B is observed. The mutual information in this case is

$$I(A, B) = \log\left(\frac{P(A)}{P(A)}\right) = \log(1) = 0 \tag{7.4}$$

On the other hand, if there is a complete dependence between A and B or in other words if observing B we are 100% sure that A has happened, then $P(A|B) = 1$ and

$$I(A, B) = \log\left(\frac{1}{P(A)}\right) = -\log(P(A)) \tag{7.5}$$

The 100% dependence assumed above is equivalent to observing A itself instead of B, hence

$$I(A, A) = -\log(P(A)) \tag{7.6}$$

which represents the maximum information we can obtain about the event A or in other words, the maximum amount of information inherent in A.

So far we have discussed single events. If we now turn to the case of having a discrete random variable X that can take one of the values x_1, x_2, ... x_K, we can define the K events as A_i when $X = x_i$. Using equation (7.6) above and taking the average of the information for all events, the **entropy** of the stochastic, discrete variable X can be calculated

$$H(X) = \mathrm{E}[I(A_i, A_i)] = -\sum_{i=1}^{K} f_X(x_i) \log\left(f_X(x_i)\right) \tag{7.7}$$

where $f_X(x_i)$ is the probability that $X = x_i$, that is, that event A_i has happened.

The entropy can be regarded as the information inherent in the variable X. As stated in the source coding theorem, this is the minimum amount of information we must keep to be able to reconstruct the behaviour of X without errors. In other words, this is the redundancy free amount of information that should preferably be produced by an ideal data compression algorithm.

An interesting thing about the entropy is that the maximum information (entropy) is obtained when all the outcomes, i.e. all the possible values x_i, are equally probable. In other words the signal X is completely random. This seems intuitively right, since when all the alternatives are equally probable the uncertainty is the greatest.

Going back to our examples in the beginning of this section, let us calculate the entropy. For the case of 'birthplace in Scandinavia' the entropy is (assuming equal probabilities)

$$H(X) = -\sum_{i=1}^{3} f_X(x_i) \, \text{lb} \left(f_X(x_i) \right) = -\frac{1}{3} \sum_{i=1}^{3} \text{lb} \left(\frac{1}{3} \right) = 1.58 \text{ bits/person}$$

Hence, the minimum amount of information for 10 independent people is 15.8 bits (we used 20 bits in our message). Data compression is therefore possible. We only have to find the 'smartest' way.

For the 'birthplace in Europe' example, the entropy can be calculated in a similar way

$$H(X) = -\sum_{i=1}^{34} f_X(x_i) \, \text{lb} \left(f_X(x_i) \right) = -\frac{1}{34} \sum_{i=1}^{3} \text{lb} \left(\frac{1}{34} \right) = 5.09 \text{ bits/person}$$

The minimum amount of information for 10 independent people is 50.9 bits; we used 60 bits in our message. Data compression is possible in this case too.

Some concluding remarks about the concept of entropy

The term entropy is commonly used in thermodynamics, where it refers to disorder of molecules. In a gas, molecules move around in random patterns, while in a crystal lattice (solid state) they are lined up in a very ordered way, in other words, there is a certain degree of dependency. The gas is said to have a higher entropy than the crystal state. Now, this has a direct coupling to the information concept. If we had to describe the position of a number of molecules, it would require more information for the gas than for the crystal. For the molecules in the gas we would have to give three coordinates (x, y, z) in space for every molecule. For the crystal, the positions could be represented by giving the coordinates for the reference corner of the lattice and the spacing of the lattice in x, y and z directions.

If you take for instance water and turn it into ice (crystal with lower entropy) you have to put the water in the freezer and add energy. Hence, **reduction of entropy requires energy**. This is also why you get exhausted by cleaning your room (reduction of entropy), while the 'messing up' process seems to be quite relaxing. In the information theory context, freezing or cleaning translates to getting rid of information, in other words 'forgetting' requires energy.

7.2 Source coding   In the previous section we concluded that getting rid of redundancy was the task of data compression. This process is also called **source coding**. The underlying idea is based on the fact that we have an **information source** generating information (discrete or continuous) at a rate of H bits/s (the entropy). If we need to store this information, we want to use as small storage as possible per time unit. If we are required to transmit the information, it is desirable to use as low data rate R as possible. From the source coding theorem, we know however that $R \geqslant H$ to avoid loss of information.

There are numerous examples of information sources, a keyboard ($+$ human) generating text, a microphone generating a speech signal, a TV camera, an image scanner, transducers in a measurement logging system etc.

The output information flow (signals) from these sources is commonly the subject of source coding (data compression) performed using general purpose digital computers or DSPs. There are two classes of data compression methods. In the first class, we find **general data compression algorithms**, that is methods that work fairly well for any type of input data (text, images, speech signals etc.). These methods are commonly able to 'decompress', in other words restore the original information data sequence without any errors ('lossless'). File compression programs for computers (for example 'double space', 'pkzip' and 'arc') are typical examples. This class of data compression methods hence adheres to the source coding theorem.

In the other class of data compression methods we find **specialized algorithms**, for instance speech coding algorithms in cellular mobile telephone systems. This kind of data compression algorithm makes use of prior information about the data sequences and may for instance contain a model of the information source. For a special type of information, these specialized methods are more efficient than the general type of algorithms. On the other hand, since they are specialized, they may exhibit poor performance if used with other types of input data. A speech-coding device designed for Nordic language speech signals may for instance be unusable for Arabian users. Another common property of the algorithms in this class is the inability to restore exactly the original information ('lossy'). They can only restore it 'sufficiently' well. Examples are speech coding systems (your mother almost sounds like your girlfriend) or image compression algorithms, producing more or less crude pictures with poor resolution and a very limited number of colour shades.

7.2.1 Huffman's algorithm

In the following discussion, we will mainly use binary digits, that is 0 and 1. Other character sets can of course be used. The basic idea of the **Huffman** source **coding** algorithm is to use variable length and prefix-free symbols and to minimize the average length of the symbols (in terms of number of characters), taking the statistics of the symbols into account. This algorithm belongs to the class of general data compression algorithms.

Variable length means that a certain symbol can be one, two, three and so on characters long, for example 0, 10, 1111 etc. To be able to decode a string of variable length symbols, no symbol is however allowed to be a prefix of a longer symbol. We must be able to identify the symbol as soon as the last character of the symbol is received. The code must be **prefix free**. A simple example of a prefix-free code is a code consisting of the symbols 0, 10, 11 while the code using the symbols 1, 11, 101 is **not** prefix free. If we for instance receive '111011' we cannot tell if it is '1', '1', '101', '1' or '11', '101', '1' or . . .?

The average length of a symbol can be expressed as

$$E[L] = \sum_{i=1}^{K} l_i f_U(u_i) \tag{7.8}$$

where l_i is the number of characters (the length) of symbol u_i and $f_U(u_i)$ is the probability of symbol u_i. The way to minimize the average symbol length (7.8) is obviously to assign the most probable symbols the shortest codes and the least probable symbols the longest codes. This idea is not new. It is for instance used in telegraphy in the form of **Morse code** (Carron, 1991). The most probable letters like 'e' and 't' have the shortest Morse codes, only one dot or one dash respectively, while uncommon characters like the question mark are coded as long sequences of dots and dashes.

Designing a code having the desired properties can be done using Huffman's algorithm, a tree algorithm. It will be demonstrated for binary codes, which result in binary trees, that is trees that branch into **two** branches, 0 and 1. Huffman's algorithm can be expressed using 'pseudo code' in the following way:

```
huffman:  assign every symbol a node
          assign every node the probability of the symbol
          make all the nodes active

          while (active nodes left) do
          {
              take the two least probable active nodes
              join these nodes into a binary tree
              deactivate the nodes
              add the probabilities of the nodes
              assign the root node this probability
              activate the root node
          }
```

The following example will be used to illustrate the way the algorithm works. Assume that we have five symbols from our information source with the following probabilities

$$u_1 \quad f_U(u_1) = 0.40$$

$$u_2 \quad f_U(u_2) = 0.20$$

$$u_3 \quad f_U(u_3) = 0.18$$

$$u_4 \quad f_U(u_4) = 0.17$$

$$u_5 \quad f_U(u_5) = 0.05$$

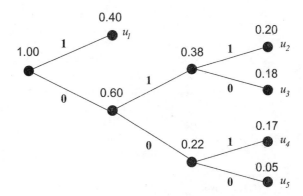

Figure 7.1 *A tree diagram used to find the optimum binary Huffman code of the example*

The resulting Huffman tree is shown in Figure 7.1 (there are many possibilities). Probabilities are shown in parentheses.

Building the tree is as follows. Firstly, we take the symbols having the smallest probability, that is u_5 and u_4. These two symbols now constitute the first little sub-tree down to the right. The sum of the probabilities is $0.05 + 0.17 = 0.22$. This probability is assigned to the root node of this first little sub-tree. The nodes u_5 and u_4 are now deactivated, that is dismissed from the further process, but the root node is activated.

Looking in our table of active symbols, we now realize that the two smallest active nodes (symbols) are u_2 and u_3. We form another binary sub-tree above the first one in a similar way. The total probability of this sub-tree is $0.20 + 0.18 = 0.38$, which is assigned to the root node.

During the next iteration of our algorithm, we find that the two active root nodes of the sub-trees are those possessing the smallest probabilities, 0.22 and 0.38 respectively. These root nodes are hence combined into a new sub-tree, having the total probability of $0.22 + 0.38 = 0.60$.

Finally, there are only two active nodes left, the root node of the sub-tree just formed and the symbol node u_1. These two nodes are joined by a last binary sub-tree and the tree diagram for the optimum Huffman code is completed. Using the definition that 'up going' (from left to right) branches are represented by a binary '1' and 'down going' branches by a '0' (this can of course be done in other ways as well), we can now write down the binary, variable length, prefix-free codes for the respective symbols:

u_1 1

u_2 011

u_3 010

u_4 001

u_5 000

Using equation (7.8) the average symbol length can be calculated, $E[L] = 2.2$ bits/symbol. The entropy can be found to be $H = 2.09$ bits/symbol using equation (7.7). As can be seen, we are quite close to the maximum data compression possible. Using fixed length binary coding would have resulted in $E[L] = 3$ bits/symbol.

Using this algorithm in the 'born in Scandinavia' example an average symbol length of $E[L] = 1.67$ bits/symbol can be achieved. The entropy is $H = 1.58$ bits/symbol. For the 'born in Europe' example the corresponding figures are $E[L] = 5.11$ bits/symbol and $H = 5.09$ bits/symbol.

The **coding efficiency** can be defined as

$$\eta = \frac{H(X)}{E[L]} \tag{7.9}$$

and the **redundancy** can be calculated using

$$r = 1 - \eta = 1 - \frac{H(X)}{E[L]} \tag{7.10}$$

If the efficiency is equal to one, the redundancy is zero and we have found the most compact, 'lossless' way of coding the information. For our examples above, we can calculate the efficiency and the redundancy

'Born in Scandinavia': $\quad \eta = \dfrac{1.58}{1.67} = 0.946 \Rightarrow r = 1 - 0.946 = 0.054$

'Born in Europe': $\quad \eta = \dfrac{5.09}{5.11} = 0.996 \Rightarrow r = 1 - 0.996 = 0.004$

In principle, the Huffman algorithm assumes that the information source is 'without memory', which means that there is no dependency between successive source symbols. The algorithm can however be used even for information sources with 'memory', but better algorithms can often be found for these cases. One such case is coding of facsimile signals, described at the end of this chapter.

7.2.2 Delta modulation (DM), ADM and CVSD

Delta modulation (DM) was briefly discussed in Chapter 2 in conjunction with sigma–delta A/D converters. The delta modulation technique can also be viewed as a data compression method. In this case, redundancy caused by dependency between successive samples in for instance a PCM data stream can be removed. The idea behind delta modulation is to use the **difference** between two successive data samples, rather than the samples themselves. Hopefully, the difference will be small and can be represented by fewer bits than the value of the samples themselves. In the simplest form of delta modulation, the difference is fixed to only ± 1 PCM step and can hence be represented by 1 bit. Assuming the PCM word has a word length of N bits, and that it is transmitted using a bit rate of R bits/s, the transmission rate

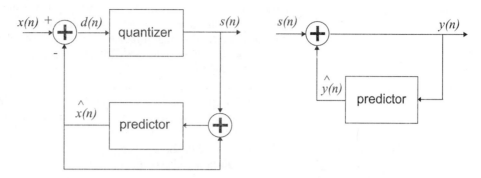

Figure 7.2 *A generic delta modulator and demodulator*

of a DM signal would be only R/N bits/s. Figure 7.2 shows a block diagram of a simple delta modulator (data compressing algorithm) and delta demodulator ('decompressing' algorithm). Starting with the modulator, the incoming data sequence $x(n)$ is compared to a predicted value $\hat{x}(n)$, based on earlier samples. The difference $d(n) = x(n) - \hat{x}(n)$ ('prediction error') is fed to a **1 bit quantizer**, in this case basically a sign detector. The output from this detector is $s(n)$

$$s(n) = \begin{cases} +\delta & \text{if } d(n) \geqslant 0 \\ -\delta & \text{if } d(n) < 0 \end{cases} \tag{7.11}$$

where δ is referred to as the **step size**, in this generic system $\delta = 1$. Further, in this basic type delta modulation system, the predictor is a **first-order predictor,** basically an accumulator, in other words a one-sample delay (memory), hence

$$\hat{x}(n) = \hat{x}(n-1) + s(n-1) \tag{7.12}$$

The signal $x(n)$ is reconstructed in the demodulator using a predictor similar to the one used in the modulator. The demodulator is fed the delta modulated signal $s(n)$, which is a sequence of $\pm\delta$ values, and the reconstructed output signal is denoted $y(n)$ in Figure 7.2

$$y(n) = \hat{y}(n) + s(n) = y(n-1) + s(n) \tag{7.13}$$

A simple DM system like the above, using a fixed step size δ is called a **linear delta modulation** system (**LDM**). There are two problems associated with this type of system, **slope overload** and **granularity**. The system is only capable of tracking a signal with a derivative (slope) smaller than $\pm\delta/\tau$ where τ is the time between two consecutive samples, that is the sampling period. If the slope of the signal is too steep, slope overload will occur. This can of course be cured by increasing the sampling rate (for instance oversampling which is used in the sigma–delta A/D converter, see Chapter 2) or by increasing the step size. By increasing the sampling rate, the perceived advantage using delta modulation may be lost. If the step size is increased, the LDM system may exhibit poor SNR performance when a weak input signal is present, due to the ripple caused by the relatively large step size. This is referred to as the granularity error.

Choosing a value of the step size δ is hence crucial to the performance of an LDM system. This kind of data compression algorithm works best for slowly varying signals having a moderate dynamic range. LDM is often inappropriate for speech signals, since in these signals there are large variations in amplitude between silent sounds (for example 'ph' or 'sh') and loud sounds (like 't' or 'a').

A way to reduce the impact of the problems above is to use **adaptive delta modulation (ADM)**. In the ADM system, the step size is adapted as a function of the output signal $s(n)$. If for instance $s(n)$ has been $+\delta$ for a certain number of consecutive samples, the step size is increased. This simple approach may improve the DM system quite a lot. However, a new problem is now present. If there are errors when transmitting the data sequence $s(n)$, the modulator and demodulator may disagree about the present step size. This may cause considerable errors that will be present for a large number of succeeding samples. A way to recover from these errors is to introduce 'leakage' into the step size adaption algorithm. Such a system is the **continuously variable slope delta modulation** system (**CVSD**).

In CVSD, the step size $\delta(n)$ depends on the two previous values of the output $s(n)$. If there have been consecutive runs of ones or zeros, there is a risk for slope overload and the step size is increased to be able to 'track' the signal faster. If on the other hand, the previous values of $s(n)$ have been alternating, the signal is obviously not changing very fast, and the step size should be reduced (by 'leakage') to minimize the granulation noise. The step size in a CVSD system is adapted as

$$\delta(n) = \begin{cases} \gamma\delta(n-1) + C_2 & \text{if } s(n) = s(n-1) = s(n-2) \\ \gamma\delta(n-1) + C_1 & \text{else} \end{cases} \qquad (7.14)$$

where $0 < \gamma < 1$ is the 'leakage' constant and $C_2 \geqslant C_1 > 0$. The constants C_1 and C_2 are used to define the minimum and maximum step size as

$$\frac{C_1}{1 - \gamma} < \delta(n) < \frac{C_2}{1 - \gamma} \qquad (7.15)$$

Still the system may suffer from slope overload and granularity problems. CVSD, being less sensitive to transmission errors than ADM for example, was however widely popular, until the advent of ADPCM (adaptive differential pulse code modulation) algorithms (see below) standardized by the CCITT (Comité Consultatif International Télégraphique et Téléphonique).

7.2.3 DPCM and ADPCM

The differential pulse code modulation (DPCM) scheme can be viewed as an extension of delta modulation presented above. The same block diagram (Figure 7.2) applies, but in the DPCM case, we have a quantizer using p values, not a 1 bit quantizer as in the former case, but rather an $\text{lb}(p)$ bit quantizer. Further, the predictor in this case is commonly of higher order, and can be described by its impulse response $h(n)$, hence

$$\hat{x}(n) = \sum_{k=1}^{N} h(k)\big(\hat{x}(n-k) + s(n-k)\big) \qquad (7.16)$$

Note, if $h(1) = 1$ and $h(k) = 0$ for $k > 1$, then we are back in the first order predictor used by the delta modulation algorithm (7.12). The predictor is commonly implemented as a FIR filter (see Chapter 1) which can be readily implemented using a DSP. The length of the filter is often quite moderate, $N = 2$ up to $N = 6$ is common.

The advantage of DPCM over straightforward PCM (see Chapter 2), is of course the data compression achieved by utilizing the dependency between samples in for instance an analog speech signal. If data compression is not the primary goal, DPCM also offers the possibility of enhancing the SNR. As DPCM is only quantizing the difference between consecutive samples and not the sample values themselves, a DPCM system has less quantization error and noise than a pure PCM system using the same word length.

To be able to find good parameters $h(n)$ for the predictor and a proper number of quantization levels p, knowledge of the signal statistics is necessary. In many cases, although the long-term statistics are known, the signal may depart significantly from these during shorter periods of time. In such applications, an adaptive algorithm may be advantageous.

Adaptive differential pulse code modulation (ADPCM) is a term used for two different methods, namely adaptation of the **quantizer** and adaptation of the **predictor** (Marven and Ewers, 1993). Adaptation of the quantizer involves estimation of the step size, based on the level of the input signal. This estimation can be done in two ways, **forward** estimation (**DPCM-AQF**) and **backward** estimation (**DPCM-AQB**) (see Figure 7.3). In the forward case, AQF, a number of input samples are first buffered in the modulator and used to estimate the signal level. This level information is sent to the demodulator, together with the normal DPCM data stream. The need to send the level information is a drawback of AQF and for this reason, AQB is more common. Another drawback of AQF is the delay introduced by the buffering process. This delay may render AQF unusable in certain applications. AQF however, has the potential of performing better than AQB.

In backward level estimation (AQB), estimation is based on the outgoing DPCM data stream. A similar estimation is made in the demodulator, using the incoming DPCM data. This scheme bears some resemblance to CVSD, discussed above. Adaptive quantizers commonly offer an improvement in SNR of roughly 3–7 dB compared to fixed quantizers. The quality of the adaptation depends to a large extent on the quality of the level estimator. There is of course a tradeoff between the complexity and the cost of a real world implementation.

When using the acronym ADPCM, adaptation of the **predictor** or adaptation of both the predictor and the quantizer is often assumed. The predictor is commonly adapted using methods based on gradient descent type algorithms used in adaptive filters (see Chapter 3). Adaptive predictors can increase the SNR of the system significantly compared to fixed predictors.

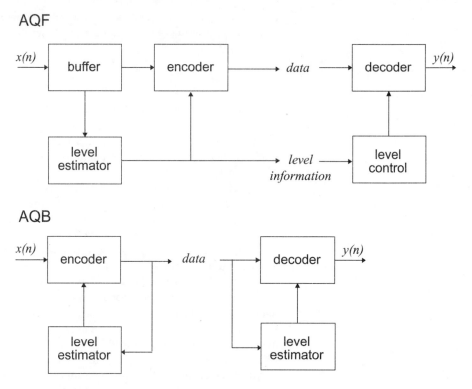

Figure 7.3 *Forward estimation (DPCM-AQF) and backward estimation (DPCM-AQB)*

As an example, in the standard CCITT G.721, the input is a CCITT standard 64 kbits/s PCM coded speech signal and the output is a 32 kbits/s ADPCM data stream. In this standard, feedback adaptation of both the quantizer and predictor is used. In a typical CCITT G.721 application (Marven and Ewers, 1993), the adaptive quantizer uses 4 bits and the predictor is made up of a 6th-order FIR filter and a 2nd-order IIR filter. A '**transcoder**' of this kind can be implemented as a mask-programmed DSP at a competitive cost.

7.2.4 Speech coding, APC and SBC

Adaptive predictive coding (APC) is a technique used for speech coding, that is data compression of speech signals. APC assumes that the input speech signal is repetitive with a period significantly longer than the average frequency content. Two predictors are used in APC. The high frequency components (up to 4 kHz) are estimated using a 'spectral' or 'formant' predictor and the low frequency components (50–200 Hz) by a 'pitch' or 'fine structure' predictor (see Figure 7.4). The spectral estimator may be of order 1–4 and the pitch estimator about order 10. The low-frequency components of the speech signal are due to the movement of the tongue, chin and

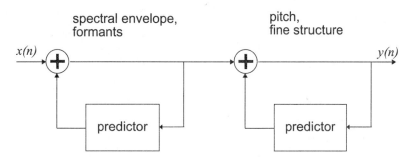

Figure 7.4 *Encoder for adaptive, predictive coding of speech signals. The decoder is mainly a mirrored version of the encoder*

lips. The high-frequency components originate from the vocal chords and the noise-like sounds (like in 's') produced in the front of the mouth.

The output signal $y(n)$ together with the predictor parameters, obtained adaptively in the encoder, are transmitted to the decoder, where the speech signal is reconstructed. The decoder has the same structure as the encoder but the predictors are not adaptive and are invoked in the reverse order. The prediction parameters are adapted for blocks of data corresponding to for instance 20 ms time periods.

APC is used for coding speech at 9.6 and 16 kbits/s. The algorithm works well in noisy environments, but unfortunately the quality of the processed speech is not as good as for other methods like CELP described below.

Another coding method is **sub-band coding (SBC)** (see Figure 7.5) which belongs to the class of **waveform coding** methods, in which the frequency domain properties of the input signal are utilized to achieve data compression.

The basic idea is that the input speech signal is split into **sub-bands** using band-pass filters. The sub-band signals are then encoded using ADPCM or other techniques. In this way, the available data transmission capacity can be allocated between bands according to perceptual criteria, enhancing the speech quality as perceived by listeners. A sub-band that is more 'important' from the human listening point of view can be allocated more bits in the data stream, while less important sub-bands will use fewer bits.

A typical setup for a sub-band coder would be a bank of N (digital) band-pass filters followed by decimators, encoders (for instance ADPCM) and a multiplexer combining the data bits coming from the sub-band channels. The output of the multiplexer is then transmitted to the sub-band decoder having a demultiplexer splitting the multiplexed data stream back into N sub-band channels. Every sub-band channel has a decoder (for instance ADPCM), followed by an interpolator and a band-pass filter. Finally, the outputs of the band-pass filters are summed and a reconstructed output signal results.

Sub-band coding is commonly used at bit rates between 9.6 kbits/s and 32 kbits/s and performs quite well. The complexity of the system may however be considerable if the number of sub-bands is large. The design of the band-pass filters is also a critical topic when working with sub-band coding systems.

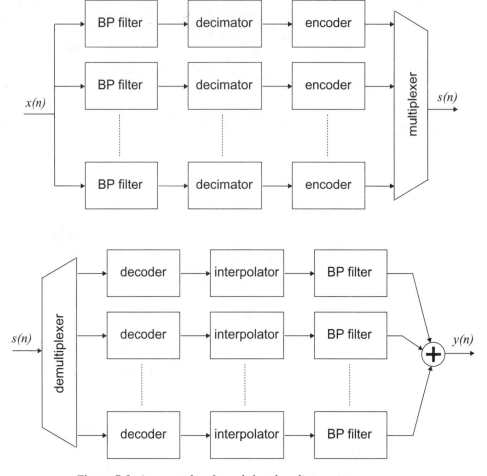

Figure 7.5 *An example of a sub-band coding system*

7.2.5 Vocoders and LPC

In the methods described above (APC, SBC and ADPCM), speech signal applications have been used as examples. By modifying the structure and parameters of the predictors and filters, the algorithms may also be used for other signal types. The main objective was to achieve a reproduction that was as faithful as possible to the original signal. Data compression was possible by removing redundancy in the time or frequency domain.

The class of **vocoders** (also called source coders) is a special class of data compression devices aimed only at speech signals. The input signal is analysed and described in terms of speech model parameters. These parameters are then used to synthesize a voice pattern having an acceptable level of perceptual quality. Hence, waveform accuracy is not the main goal, as in the previous methods discussed.

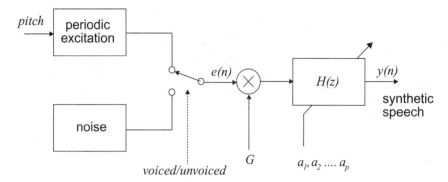

Figure 7.6 *The LPC model*

The first vocoder was designed by H. Dudley in the 1930s and demon-strated at the 'New York Fair' in 1939. Vocoders have become popular as they achieve reasonably good speech quality at low data rates, from 2.4 kbits/s to 9.6 kbits/s. There are many types of vocoders (Marven and Ewers, 1993), some of the most common techniques will be briefly presented below.

Most vocoders rely on a few basic principles. Firstly, the characteristics of the speech signal is assumed to be fairly constant over a time of approx-imately 20 ms, hence most signal processing is performed on (overlapping) data blocks of 20–40 ms length. Secondly, the speech model consists of a time varying filter corresponding to the acoustic properties of the mouth and an excitation signal. The excitation signal is either a periodic waveform, as created by the vocal chords, or a random noise signal for production of 'unvoiced' sounds, for example 's' and 'f'. The filter parameters and exci-tation parameters are assumed to be independent of each other and are commonly coded separately.

Linear predictive coding (LPC) is a popular method, which has however been replaced by newer approaches in many applications. LPC works exceed-ingly well at low bit rates and the LPC parameters contain sufficient information of the speech signal to be used in speech recognition applica-tions. The LPC model is shown in Figure 7.6.

LPC is basically an **auto-regressive** model (see Chapter 5) and the vocal tract is modelled as a time-varying all-pole filter (IIR filter) having the transfer function $H(z)$

$$H(z) = \frac{1}{1 + \sum_{k=1}^{p} a_k z^{-k}} \tag{7.17}$$

where p is the order of the filter. The excitation signal $e(n)$, being either noise or a periodic waveform, is fed to the filter via a variable gain factor G. The output signal can be expressed in the time domain as

$$y(n) = Ge(n) - a_1 y(n-1) - a_2 y(n-2) - \ldots - a_p y(n-p) \tag{7.18}$$

The output sample at time n is a linear combination of p previous samples and the excitation signal (linear predictive coding). The filter coefficients a_k are time varying.

The model above describes how to **synthesize** the speech given the pitch information (if noise or periodic excitation should be used), the gain and the filter parameters. These parameters must be determined by the encoder or the **analyser**, taking the original speech signal $x(n)$ as input.

The analyser windows the speech signal in blocks of 20–40 ms, usually with a Hamming window (see Chapter 5). These blocks or 'frames' are repeated every 10–30 ms, hence there is a certain overlap in time. Every frame is then analysed with respect to the parameters mentioned above.

Firstly, the pitch frequency is determined. This also tells whether we are dealing with a voiced or unvoiced speech signal. This is a crucial part of the system and many pitch detection algorithms have been proposed. If the segment of the speech signal is voiced and has a clear periodicity or if it is unvoiced and not periodic, things are quite easy. Segments having properties in between these two extremes are difficult to analyse. No algorithm has been found so far that is 'perfect' for all listeners.

Now, the second step of the analyser is to determine the gain and the filter parameters. This is done by estimating the speech signal using an adaptive predictor. The predictor has the same structure and order as the filter in the synthesizer. Hence, the output of the predictor is

$$\hat{x}(n) = -a_1 x(n-1) - a_2 x(n-2) - \ldots - a_p x(n-p) \tag{7.19}$$

where $\hat{x}(n)$ is the predicted input speech signal and $x(n)$ is the actual input signal. The filter coefficients a_k are determined by minimizing the square error

$$\sum_n \left(x(n) - \hat{x}(n) \right)^2 = \sum_n r^2(n) \tag{7.20}$$

This can be done in different ways, either by calculating the auto-correlation coefficients and solving the Yule–Walker equations (see Chapter 5) or by using some recursive, adaptive filter approach (see Chapter 3).

So, for every frame, all the parameters above are determined and transmitted to the synthesizer, where a synthetic copy of the speech is generated.

An improved version of LPC is **residual excited linear prediction (RELP)**. Let us take a closer look at the error or residual signal $r(n)$ resulting from the prediction in the analyser (equation (7.19)). The residual signal (we are trying to minimize) can be expressed as

$$r(n) = x(n) - \hat{x}(n)$$

$$= x(n) + a_1 x(n-1) + a_2 x(n-2) + \ldots + a_p x(n-p) \tag{7.21}$$

From this it is straightforward to find out that the corresponding expression using the z-transforms is

$$R(z) = \frac{X(z)}{H(z)} = X(z) H^{-1}(z) \tag{7.22}$$

Hence, the predictor can be regarded as an 'inverse' filter to the LPC model filter. If we now pass this residual signal to the synthesizer and use it to excite the LPC filter, that is $E(z) = R(z)$, instead of using the noise or periodic waveform sources we get

$$Y(z) = E(z)H(z) = R(z)H(z) = X(z)H^{-1}(z)H(z) = X(z) \qquad (7.23)$$

In the ideal case, we would hence get the original speech signal back. When minimizing the variance of the residual signal (equation (7.20)), we gathered as much information about the speech signal as possible using this model in the filter coefficients a_k. The residual signal contains the remaining information. If the model is well suited for the signal type (speech signal), the residual signal is close to white noise, having a flat spectrum. In such a case we can get away with coding only a small range of frequencies, for instance 0–1 kHz of the residual signal. At the synthesizer, this baseband is then repeated to generate higher frequencies. This signal is used to excite the LPC filter.

Vocoders using RELP are used with transmission rates of 9.6 kbits/s. The advantage of RELP is a better speech quality compared to LPC for the same bit rate. However, the implementation is more computationally demanding.

Another possible extension of the original LPC approach is to use **multi-pulse excited linear predictive coding (MLPC)**. This extension is an attempt to make the synthesized speech less 'mechanical', by using a number of different pitches of the excitation pulses rather than only the two (periodic and noise) used by standard LPC.

The MLPC algorithm sequentially detects k pitches in a speech signal. As soon as one pitch is found it is subtracted from the signal and detection starts over again, looking for the next pitch. Pitch information detection is a hard task and the complexity of the required algorithms is often considerable. MLPC however offers a better speech quality than LPC for a given bit rate and is used in systems working with 4.8–9.6 kbits/s.

Yet another extension of LPC is the **code excited linear prediction (CELP)**. The main feature of the CELP compared to LPC is the way in which the filter coefficients are handled. Assume that we have a standard LPC system, with a filter of the order p. If every coefficient a_k requires N bits, we need to transmit $N \cdot p$ bits per frame for the filter parameters only. This approach is all right if all combinations of filter coefficients are equally probable. This is however not the case. Some combinations of coefficients are very probable, while others may never occur. In CELP, the coefficient combinations are represented by p dimensional vectors. Using vector quantization techniques, the most probable vectors are determined. Each of these vectors are assigned an **index** and stored in a **codebook**. Both the analyser and synthesizer of course have identical copies of the codebook, typically containing 256–512 vectors. Hence, instead of transmitting $N \cdot p$ bits per frame for the filter parameters only 8–9 bits are needed.

This method offers high-quality speech at low-bit rates but requires considerable computing power to be able to store and match the incoming speech to the 'standard' sounds stored in the codebook. This is of course especially true if the codebook is large. Speech quality degrades as the codebook size decreases.

Most CELP systems do not perform well with respect to higher frequency components of the speech signal at low bit rates. This is counteracted in newer systems using a combination of CELP and MLPC.

There is also a variant of CELP called **vector sum excited linear prediction (VSELP)**. The main difference between CELP and VSELP is the way the codebook is organized. Further, since VSELP uses fixed point arithmetic algorithms, it is possible to implement using cheaper DSP chips than CELP, which commonly requires floating point arithmetics.

7.2.6 Image coding, JPEG and MPEG

Digitized images consist of large amounts of picture elements (pixels), hence transmitting or storing images often involves large amounts of data. For this reason, data compression of image data, or image coding is a highly interesting topic.

The same general fundamental idea for data compression applies in that there is a reduction of redundancy utilizing statistical properties of the data set such as dependencies between pixels. In the case of image compression, many of the algorithms turn two dimensional, unless for instance compression is applied to a scanned signal. One such application is telefax (facsimile) systems, that is systems for transmission (or storage) of black-and-white drawings or maps etc. In the simplest case, there are only two levels to transmit, black and white, which may be represented by binary '1' and '0' respectively, coming from the scanning photoelectric device.

Studying the string of binary digits (symbols) coming from the scanning device (the information source), it is easy to see that ones and zeros often come in long **bursts**, and not in a 'random' fashion; for instance 30 zeros, followed by 10 ones, followed by 80 zeros and so on may be common. There is hence a considerable dependency between successive pixels. In this case, a **run length code** will be efficient. The underlying idea is to not transmit every single bit, but rather a number telling how many consecutive ones or zeros there are in a burst.

For facsimile applications, CCITT has standardized a number of run length codes. A simple code is the **Modified Huffman Code (MHC).** Black and white bursts of length 0–63 pixels are assigned their own code words according to the Huffman algorithm outlined earlier in this chapter. If a burst is 64 pixels or longer, the code word is preceded by a prefix code word, telling how many multiples of 64 pixels there are in the burst. The fact that long white bursts are more probable than long black bursts is also taken into account when assigning code words. Since the letters are 'hollow', a typical page of text contains approximately 10% black and 90% white areas.

During the scanning process, every scan line is assumed to start with a white pixel, else the code word for white burst length zero is transmitted. Every scanned line is ended by a special 'end-of-line' code word to reduce the impact of possible transmission errors. Compression factors between 6 and 16 times can be observed (that is 0.06–0.17 bits/pixel), depending on the appearance of the scanned document. A more elaborate run length code is the **modified READ code (MRC)** (Hunter and Robinson, 1980) capable of higher compression factors.

Image compression algorithms of the type briefly described above are often denoted **lossless** compression techniques, since they make a faithful reproduction of the image after decompression, that is no information is lost. If we can accept a certain degradation of for instance resolution or contrast in a picture, different schemes of predictive compression and transform compression can be used.

Predictive compression techniques work in similar ways as DPCM and ADPCM described earlier. The predictors may however be more elaborate, working in two dimensions and utilizing dependencies in both the x and y direction of an image. For moving pictures, predictors may also consider dependencies in the z axis, between consecutive picture frames. This is sometimes called three-dimensional prediction. For slowly moving objects in a picture, the correlation between picture frames may be considerable, offering great possibilities for data compression. One such example is the videophone. The most common image in such a system is the more or less static face of a human.

The basic idea of **transform compression** (Gonzalez and Wintz, 1987) is to extract appropriate statistical properties, for instance Fourier coefficients, of an image and let the most significant of these properties represent the image. The image is then reconstructed (decompressed) using an inverse transform.

A picture can be Fourier transformed in a similar way to a temporal one-dimensional signal, and a spectrum can be calculated. Dealing with images however, we have two dimensions and hence two frequencies, one in the x direction and one in the y direction. Further, we are now dealing with **spatial** frequencies. For instance a coarse chess board pattern would have a lower spatial frequency than a fine pattern. We are now talking about cycles per length unit, not cycles per time unit.

Often it is convenient to express the transform coefficients as a matrix. In doing this, it is commonly found that the high-frequency components have smaller amplitude than lower frequency components. Hence, only a sub-set of the coefficients need to be used to reproduce a recognizable image, but fine structure details will be lost. Transform compression methods are commonly not lossless.

It has been found that standard two-dimensional FFT is not the best choice for image compression. Alternative transforms, for instance **Walsh**, **Hadamard** and **Karhunen–Loève** transforms have been devised. One of the most popular is however the two-dimensional **discrete cosine transform (DCT)**

$$C(0, 0) = \frac{1}{N} \sum_{x=0}^{N-1} \sum_{y=0}^{N-1} f(x, y) \qquad (7.24a)$$

and for $u, v = 1, 2, \ldots N-1$

$$C(u, v) = \frac{1}{2N^3} \sum_{x=0}^{N-1} \sum_{y=0}^{N-1} f(x, y)\cos\big((2x+1)u\pi\big)\cos\big((2y+1)v\pi\big) \qquad (7.24b)$$

where $C(u, v)$ is the two-dimensional transform coefficient ('spectrum'), $C(0, 0)$ is the 'DC' component and $f(x, y)$ is the pixel value. The inverse transform is

$$f(x, y) = \frac{1}{N} C(0, 0)$$

$$+ \frac{1}{2N^3} \sum_{u=0}^{N-1} \sum_{v=0}^{N-1} C(u, v)\cos((2x+1)u\pi)\cos((2y+1)v\pi) \qquad (7.25)$$

One nice feature of DCT is that the transform is real, unlike FFT which is a complex transform. Another advantage is that DCT is separable, implying that it can be implemented as two successive applications of a one-dimensional DCT algorithm.

A common technique is to spilt a picture into 8 x 8 pixel blocks and apply DCT. High-amplitude, low-frequency coefficients are transmitted first. In this way, a 'rough' picture is first obtained with an increasing resolution as higher frequency coefficients arrive. Transform coding can offer a data compression ratio of approximately 10 times.

The CCITT H.261 standard covers a class of image compression algorithms for transmission bit rates from 64 kbits/s to 2 Mbits/s. The lowest bit rate can be used for videophones on narrowband ISDN lines. The H.261 is a hybrid DPCM/DCT system with motion compensation. The luminance (black-and-white brightness) signal is sampled at 6.75 MHz, and the chrominance (colour information) signal is sampled at 3.375 MHz. The difference between the present frame and the previous one is calculated and split into 8 x 8 pixel blocks. These blocks are transformed using DCT, and the resulting coefficients are coded using Huffman coding.

The motion detection algorithm takes each 8×8 pixel block of the present frame and searches the previous frame by moving the block ±15 pixels in the x and y directions. The best match is represented by a displacement vector. The DCT coefficients and the displacement vector are transmitted to the decompressor, where the reverse action takes place.

There are commercial vidoeconferencing systems available today which are using the H.261 standard. The algorithms used are however very computationally demanding. Floating-point or multiple fixed-point DSPs are required.

The **joint photographics expert group (JPEG)** is a proposed standard for compression of still pictures. The colour signals red, green and blue are sampled and each colour component is transformed by DCT in 8×8 pixel blocks. The DCT coefficients are quantized and encoded in a way that the more important lower frequency components are represented by more bits than the higher frequency coefficients. The coefficients are reordered by reading the DCT coefficient matrix in a zigzag fashion (Marven and Ewers, 1993), and the data stream is Huffman coded (the 'DC' component is differentially encoded with the previous frame, if there were any).

The JPEG compressor is simpler than in the H.261 system. There is for instance no motion compensation, but many other elements are quite similar. The decompressor is however more complicated in JPEG than in H.261.

The **moving pictures expert group (MPEG)** proposed standard (MPEG 1) is aimed for compression of full-motion pictures on digital storage media, for instance CD-ROM and DVD with a bit transfer rate of about 1.5 Mbits/s. It is to some extent similar to both H.261 and JPEG but does not have the motion compensation found in JPEG.

A sampled frame is spilt into blocks and transformed using DCT in the same way as for JPEG. The coefficients are then coded with either forward or backward prediction or a combination of both. The output from the predictive coding is then quantized using a matrix of quantization steps. Since MPEG is more complicated than JPEG it requires even more computing power.

The area of image and video compression algorithms is constantly evolving and there are many new methods and novel, dedicated signal processing ASICs and DSPs to come. Future image compression techniques may include fractal and wavelet-based methods.

7.2.7 The Lempel–Ziv algorithm

There are a number of data compression algorithms based on the **Lempel–Ziv (LZ)** method, named after A. Lempel and J. Ziv. The original ideas were published in 1977 and 1978. Only the basic method will be briefly described below.

The LZ data compression algorithm is a called a 'universal' algorithm since it does not need any prior information about the input data statistics. Commonly, it works on a byte basis and is best suited for compression of e.g. ASCII (American standard code for information interchange) information. The algorithm creates a table or 'dictionary' of frequently seen strings of bytes. When a string appears in the input data set, it is substituted by the index of a matching string in the dictionary. This dictionary type algorithm is also referred to as a macro-replacement algorithm, because it replaces a string with a token. Decompression will simply be a table-lookup and string replacement operation. It is imperative however that no transmission errors occur, since these may add errors to the dictionary and jeopardize the entire decompression operation.

Let us illustrate the algorithm with a simple example. Assume that we are using an 8 bit ASCII code and want to transmit the following string

```
the_theme_theorem_theses
```

Since a space _ also represents an ASCII character, we have to transmit 24 characters, that is a total of 24 x 8 = 192 bits.

Now, let us use a simple version of the LZ method and limit the size of the dictionary to 16 entries, so any entry in the dictionary can be pointed to by a 4 bit index. The compression and dictionary handling rule works like this:

```
lzcomp:     clear dictionary
            start from beginning of string
            while (string not traversed)
            {
```

```
find the longest substring matching an
  entry in dictionary
get the index of the entry in the
  dictionary
get the next character in the string
  following the substring
transmit index and next character
append the substring + next character to
  the dictionary
continue after next character in string
}
```

The corresponding decompression and dictionary handling rule is:

```
lzdecomp:   clear dictionary
            clear buffer for decompressed string
            while (compressed characters received)
            {
                get index and next character from
                  compressor
                use index to fetch substring from
                  dictionary
                append substring to buffer
                append next character to buffer
                append the substring + next character to
                  the dictionary
            }
```

In Table 7.1 the string, dictionary and transmitted symbols are shown. The indices are represented by the hexadecimal digits: 0, 1, . . . E, F.

The uncompressed string:

```
the_theme_theorem_theses
```

When the algorithm is started, the dictionary is empty. The first entry will be 'nothing' which will be given index 0. The first sub-string in the input string, simply the letter 't' cannot be found in the dictionary. However, 'nothing' can be found so we concatenate 'nothing' and 't' (we do not include 'nothing' in the dictionary from now on, since it is so small . . .) and add it to the dictionary. The index of this new entry will be 1. We also submit the hexadecimal sub-string index 0 ('nothing') and the letter 't' to the output. These steps are repeated and the dictionary grows. Nothing interesting happens until we find the sub-string 't' for the second time (in 'theme'). Since the substring 't' is already in the dictionary, it can be replaced by its index 1. So, following the rules, we send the index 1 and the next character 'h', further we add the new sub-string 'th' to the dictionary. This process is further repeated until the end of the uncompressed input string is reached. The total output compressed string will be:

Table 7.1 *Example LZ, 16 entry dictionary*

Index	Dictionary	Compressed output
0	nothing	
1	t	0t
2	h	0h
3	e	0e
4	_	0_
5	th	1h
6	em	3m
7	e_	3_
8	the	5e
9	o	0o
A	r	0r
B	em_	6_
C	thes	8s
D	es	3s
E		
F		

0t0h0e0_1h3m3_5e0o0r6_8s3s

where every second character is an index (4 bits) and an ASCII character (8 bits) respectively. The total length of the compressed string will be: $13 \times 4 + 13 \times 8 = 156$ bits, implying a compression factor of $156/192 = 0.81$.

The decompressor receives the compressed string above and starts building the decompressed string and the dictionary in a similar way. To start with 'nothing' is assumed in position 0 of the dictionary. The first compressed information is index 0, corresponding to 'nothing' in the dictionary and the character 't'. Hence 'nothing' and 't' are placed in the first position of the decompressed string and the same information is added to the dictionary at position 1. The process is then repeated until the end of the compressed string. Now the decompressed string is an exact copy of the uncompressed string and the dictionary is an exact copy of the dictionary used at compression time. It is worth noting that there are also many variations on this popular data compression algorithm.

8 Error-correcting codes

8.1 Channel coding

In Chapter 7, the source coding theorem of Shannon was briefly presented. In this chapter, some implications of his **channel coding theorem** will be discussed. The idea of source coding or data compression is to find a 'minimum form' in which to represent a certain amount of information, thus making transmission faster and/or data storing more compact. Data compression is based on removing redundancy, that is 'excess' information. A problem however arises; the redundancy-free pieces of information will be extremely **vulnerable**. One example of this is the data compression algorithms used to 'compress' data files on magnetic discs. If one single bit in the 'wrong' place happens to be corrupt, the entire disc (maybe several gigabytes) may be completely unreadable and almost all the information lost.

The same problem is even more likely to appear when transmitting data over wires or radio links. In such applications, transmission errors are common. Hence, to be able to transmit information error free, we must include mechanisms capable of detecting and preferably correcting transmission (or storage) errors. This requires **adding redundancy**. When adding redundancy, that is making a compressed piece of data large again, we take the properties of the transmission (or storage) process into account. If we know what kinds of errors are the most probable, for instance in a radio link, we can add redundant information tailored to correct these errors. **Channel coding** is mainly to 'vaccinating' the information against expected transmission errors. If unexpected errors occur, we will of course be in trouble. If we however design our channel code properly, this will rarely occur.

As mentioned above, adding this 'error-correcting' redundancy will increase the amount of data involved. Hence, we are making a tradeoff between transmission time or storage space for reliability.

8.1.1 The channel model

In the following text, we will denote both a data transmission (in space) and a data storing (transmission in time) mechanism a '**channel**'. This is hence an abstract entity and will not necessarily relate to any physical media, device or system. It will only be used as a model, that is a description of how the received (or retrieved) data is related to the transmitted (or stored) data.

In Chapter 7, the concept of mutual information $I(A, B)$ was introduced (equation (7.2))

$$I(A, B) = \log\left(\frac{P(A|B)}{P(A)}\right)$$

(8.1)

where $P(A)$ is the probability that event A, generated by an information source, takes place. $P(A|B)$ is the conditional probability of A taking place given we have observed another event B. Now, in the channel context, event A corresponds to the transmitted data symbol and event B to the received symbol, in other words what we observe is probably caused by A.

Now, for a simple example, let us assume that binary digits 1 and 0 are used to communicate over a channel and that the input symbols of the channel are denoted X and the output symbols Y. Event A_i implies that $X = x_i$ and event B_j implies that $Y = y_j$. We can rewrite (equation (8.1)) as

$$I(X = x_i, Y = y_i) = \log\left(\frac{P(X = x_i | Y = y_j)}{P(X = x_i)}\right) \tag{8.2}$$

We now define the **average mutual information**

$$I(X, Y) = \mathrm{E}[I(A_i, B_j)] = \sum_{i=1}^{K}\sum_{j=1}^{L} f_{XY}(x_i, y_j) \log\left(\frac{f_{X|Y}(x_i|y_j)}{f_X(x_i)}\right)$$

$$= \sum_{i=1}^{K}\sum_{j=1}^{L} f_{XY}(x_i, y_j) \log\left(f_{X|Y}(x_i|y_j)\right) - \sum_{i=1}^{K} f_X(x_i) \log\left(f_X(x_i)\right)$$

$$= -H(X|Y) + H(X) \tag{8.3}$$

where $H(X)$ is the entropy as defined in the previous chapter (equation 7.7) or the 'input information'. Further, we have defined the conditional entropy $H(X|Y)$ in a similar way, which can be viewed as the 'uncertainty' caused by the channel.

In Figure 8.1 a simple but useful model called the **binary symmetric channel (BSC)** is shown. Symbols 1 or 0 are input from the left, and output to the right. The BSC is 'memoryless', which means that there is no dependency between successive symbols. The numbers on the branches are the corresponding **transition probabilities**, that is the conditional probabilities $P(Y|X)$ which in this case can be expressed in terms of the error probability p

$$P(Y = 1 | X = 1) = (1 - p)$$

$$P(Y = 1 | X = 0) = p$$

$$P(Y = 0 | X = 1) = p$$

$$P(Y = 0 | X = 0) = (1 - p)$$

As shown in equation (7.3) the mutual information is symmetric with respect to X and Y, hence using Bayes' theorem equation (8.3) can be rewritten as

$$I(X, Y) = \mathrm{E}[I(A_i, B_j)] = \mathrm{E}[I(B_j, A_i)]$$

$$= \sum_{j=1}^{L}\sum_{i=1}^{K} f_{XY}(x_i, y_j) \log\left(\frac{f_{Y|X}(y_j|x_i)}{f_Y(y_j)}\right)$$

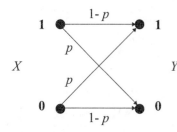

Figure 8.1 *The binary symmetric channel (BSC), error probability p*

$$= \sum_{j=1}^{L} \sum_{i=1}^{K} f_{XY}(x_i, y_j) \log \left(f_{Y|X}(y_j|x_i) \right) - \sum_{j=1}^{L} f_Y(y_j) \log \left(f_Y(y_j) \right)$$

$$= \sum_{j=1}^{L} \sum_{i=1}^{K} f_{Y|X}(y_j|x_i) f_X(x_i) \log \left(f_{Y|X}(y_j|x_i) \right) - \sum_{j=1}^{L} f_Y(y_j) \log \left(f_Y(y_j) \right)$$

$$= -H(Y|X) + H(Y) \tag{8.4}$$

where the probabilities

$$f_Y(y_j) = \sum_{i=1}^{K} f_{Y|X}(y_j|x_i) f_X(x_i)$$

Assume that the inputs $x_1 = 0$ and $x_2 = 1$ are equally probable i.e. $f_X(0) = f_X(1) = 0.5$. This implies an entropy (equation (7.7)) of (using logarithm base 2)

$$H(X) = -\sum_{i=1}^{2} f_X(x_i) \, \mathrm{lb}(f_X(x_i)) = -\frac{1}{2} \mathrm{lb} \left(\frac{1}{2} \right) - \frac{1}{2} \mathrm{lb} \left(\frac{1}{2} \right) = 1 \text{ bit/symbol}$$

Not very surprisingly, this is the **maximum** entropy we can get from one binary symbol (a bit). If for instance $x_1 = 0$ is more probable than $x_2 = 1$ or vice versa, then $H(X) < 1$ bit.

For a 'perfect' channel, that is a channel that does not introduce any errors or the error probability, $p = 0$. This implies that

$$P(Y = 1 | X = 1) = 1$$

$$P(Y = 1 | X = 0) = 0$$

$$P(Y = 0 | X = 1) = 0$$

$$P(Y = 0 | X = 0) = 1$$

Inserting the conditional probabilities into equation (8.4) we obtain

$$I(X, Y) = H(Y) - H(Y|X) = H(Y) = 1 \text{ bit/symbol} \tag{8.5}$$

The conditional entropy, or in other words the uncertainty of the channel $H(Y|X) = 0$ and all the 'input entropy' ('information') are obtained on the output of the channel. By observing Y we know everything about X.

If a BSC has an error probability of $p = 0.1$, on average 10% of the bits are erroneous, which corresponds to 10% of the bits being inverted. The transition probabilities can readily be calculated

$$P(Y = 1 | X = 1) = 0.9$$

$$P(Y = 1 | X = 0) = 0.1$$

$$P(Y = 0 | X = 1) = 0.1$$

$$P(Y = 0 | X = 0) = 0.9$$

Again, using (equation (8.4)) the resulting average mutual information will be

$$I(X, Y) = H(Y) - H(Y|X) = 1 - 0.469 = 0.531 \text{ bit/symbol} \qquad (8.6)$$

In this case, we only gain 0.531 bits of information about X by receiving the symbols Y. The uncertainty in the channel has 'diluted' the information about the input symbols X. Finally, for the worst channel possible, the error probability is $p = 0.5$. On the average we get errors half of the time. This **is** the worst case, since if p is larger than 0.5, that is the channel inverts more than half the number of transmitted bits, we could simply invert Y and thus reduce the error probability. The transition probabilities are

$$P(Y = 1 | X = 1) = 0.5$$

$$P(Y = 1 | X = 0) = 0.5$$

$$P(Y = 0 | X = 1) = 0.5$$

$$P(Y = 0 | X = 0) = 0.5$$

Inserting into (equation (8.4)) we get

$$I(X, Y) = H(Y) - H(Y|X) = 1 - 1 = 0 \text{ bit/symbol} \qquad (8.7)$$

Such a result can be interpreted as if no information about X progresses through the channel. This is due to the great uncertainty in the channel itself. In other words, whether we obtain our data sequence Y from a random generator or from the output of the channel does not matter. We would obtain equal amounts of knowledge about the input data X in both cases.

The BSC is a simple channel, however, in a realistic case more complicated models often have to be used. These models may be non-symmetric and use M number of symbols rather than only 2 (0 and 1). An 'analog' channel (continuous in amplitude) can be regarded as having an infinite number of symbols. Further, channel models can possess memory, introducing dependency between successive symbols.

8.1.2 The channel capacity

From the above discussion it is clear that every channel has a limited ability to transfer information (average mutual information), and that this limit depends on the transition probabilities, that is the error probability. The maximum average mutual information possible for a channel is called **the channel capacity**. For the discrete, memoryless channel, the channel capacity is defined as the following

$$C = \sup_{f_X(x)} I(X, Y) \qquad (8.8)$$

By changing the probability density distribution $f_X(x)$ of the symbols transmitted over the channel or by finding a proper **channel code**, the average mutual information can be maximized. This maximum value, the channel capacity, is a property of the channel and cannot be exceeded if we want to transmit information with an arbitrarily low error rate. This is mainly what the channel coding theorem is about.

It is important to note that regardless of the complexity of the devices and algorithms, if we try to exceed the channel capacity by transmitting more information per time unit (per symbol) than the channel capacity allows, we will never be able to transmit the information error free. Hence, the channel capacity can be regarded a universal 'speed limit'. If we try to exceed this information transfer limit, we will be 'fined' in terms of errors, and the net amount of transmitted error-free information will never exceed the channel capacity.

Unfortunately, the channel coding theorem does not give any hint on how to design the optimum channel code needed to obtain the maximum information transfer capacity, that is the channel capacity. The theorem only implies that there exists at least one such channel code. The search for the most effective channel codes has been pursued during the last 40 years, and some of these results will be presented later in this chapter.

Calculating the channel capacity for channels in 'reality' is in many cases very hard or even impossible. Two simple, common, standard channels should however be mentioned, the BSC and the AWGN (additive white Gaussian noise) channel.

Firstly, the BSC channel capacity can be represented as follows (Figure 8.1)

$$C = 1 + p\mathrm{lb}(p) + (1 - p)\mathrm{lb}(1 - p) \text{ bit/symbol} \qquad (8.9)$$

Secondly, the common memoryless channel model for 'analog' (continuous) signals, the **AWGN channel**, shown in Figure 8.2, is slightly more elaborate. The analog input signal is X and the output signal is Y. N is a white Gaussian noise signal, added to the input signal. The term 'white' is used in the sense that the noise has equal spectral density at all frequencies, similar to white light that has equal power at all wavelengths in the visible spectrum (contains equal amounts of all colours).

In the context of the AWGN channel, the **signal-to-noise ratio (SNR)** is commonly used, simply being the ratio between the signal power and the noise power (Ahlin and Zander, 1998)

$$SNR = \frac{X}{N} \qquad (8.10)$$

The SNR is often expressed in decibels (dB)

$$SNR_{dB} = 10 \lg (SNR) \qquad (8.11)$$

where lg ( ) is the logarithm base 10. Thus the channel capacity of the AWGN channel can be shown to be (Cover and Thomas, 1991)

$$C = W \mathrm{lb} (1 + SNR) = W \mathrm{lb} \left(1 + \frac{X}{N}\right) = W \mathrm{lb} \left(1 + \frac{X}{WN_0}\right) \text{ bit/s} \qquad (8.12)$$

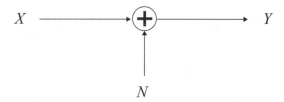

$X \longrightarrow \oplus \longrightarrow Y$

N

Figure 8.2 *The additive white Gaussian noise channel (AWGN)*

where W is the bandwidth and N_0 is the spectral density of the noise in W/Hz. An interesting example is a common telephone subscriber loop. If we assume a bandwidth of $W = 4$ kHz and $SNR = 45$ dB, the resulting channel capacity is (approximately) $C = 60$ kbit/s. Hence, the 57600 baud dial-up modems used today are approaching the theoretical upper limit. Some subscriber loops may of course have other bandwidth and SNR characteristics, which will affect the channel capacity.

Finally, to conclude this section, assume we have an information source with entropy H bit/s. In the previous chapter, the source coding theorem stated that to be able to reconstruct the information without errors, we must use an information rate R of $R \geqslant H$ bit/s. This is hence a data compression limit. In this chapter, the channel coding theorem implies that to be able to transmit information error free at information rate R, $C \geqslant R$ bit/s. This is a communication speed limit.

8.2 Error-correcting codes

There are a number of different error-correcting codes. These codes can be divided into classes, depending on their properties. A coarse classification is to divide the codes into two groups: **block codes** and **convolution codes**.

In this section, we will generally discuss coding and properties of codes and follow this with a presentation of some common block and convolution code algorithms.

8.2.1 Hamming distance and error correction

We shall begin with a brief illustration of a crude yet simple **repetition code**. Assume that we need to transmit information about a binary variable U. The probabilities of $u = 0$ and $u = 1$ are equal, which means $f_U(u{=}0) = f_U(u{=}1) = 0.5$. When transmitting this sequence of ones and zeros over a BSC with an error probability of $p = 0.1$, having a channel capacity $C = 0.531$ bit/symbol (equation (8.6)), we will get 10% bit errors on the average.

By adding redundancy for instance if we intend to transmit a '1', we repeat it twice. In other words, we transmit '111' and if we want to send a zero, we send the channel code word '000'. If we get an error in one bit of a code word, we can still correct the error by making a majority vote when receiving the code. If there are more ones than zeros in a received code word, the information symbol sent was probably $u = 1$, and if there are more zeros than ones, $u = 0$ is the most likely alternative. This is simply called the **maximum likelihood (ML)** principle. The code word X is structured in the way

$$X = \begin{bmatrix} x_1 & x_2 & x_3 \end{bmatrix} = \begin{bmatrix} u & p_1 & p_2 \end{bmatrix} \tag{8.13}$$

Where u is the information bit, and p_1 and p_2 are denoted **parity bits** or **check bits**. If a code word is structured in such a way that the parity bits and information bits are in separate groups rather than a mixed order, the code word is said to have a **systematic form**. The coding process in this example will be very easy

$$x_1 = u$$

$$x_2 = p_1 = u \tag{8.14}$$

$$x_3 = p_2 = u$$

The decoding process will also be easy, a pure majority vote as outlined above. The length of the code word is often denoted by n and the number of information bits k, hence in our example above, $n = 3$ and $k = 1$. Now, the **code rate** or code **speed** is obtained from

$$R = \frac{k}{n} \tag{8.15}$$

This can be seen as a 'mix' ratio between the numbers of necessary information bits and the total number of bits in the code word (information bits + parity bits). In the first case above, transmitting U directly, without using any code, implies a rate of

$$R = \frac{1}{1} = 1 > C = 0.531$$

In this case, we have violated the channel coding theorem and consequently the length of the strings of symbols we send does not matter, as $n \rightarrow \infty$ we will still not be able to reduce the average error rate to below 10%.

In the latter case, using the repetition code we have introduced redundancy and for this case

$$R = \frac{1}{3} = 0.333 < C = 0.531$$

It can be shown that for longer code words, that is as $n \rightarrow \infty$ the average error rate will tend to zero. Let us take a closer look at the underlying processes. Table 8.1 shows the eight possible cases for code word transmission and the associated transition probabilities expressed in terms of the error probability p of the BSC. From Table 8.1, in case (a) we intended to transmit $u = 1$ which was coded as 111. No errors occurred which has the probability of $(1 - p)^3$ $= (1 - 0.1)^3 = 0.729$ and 111 was received. This received code word was decoded using a majority vote and found to be $\hat{u} = 1$. In case (b) a single bit error occurred and hence one bit in the received code word is corrupt. This

Table 8.1 Transmission of simple 3 bit repetition code

	X Y	Probability	
(a)	$111 \rightarrow 111$	$(1-p)^3$	no errors
(b)	$111 \rightarrow 110, 101, 011$	$3(1-p)^2p$	1 bit error
(c)	$111 \rightarrow 001, 010, 100$	$3(1-p)p^2$	2 bit error
(d)	$111 \rightarrow 000$	p^3	3 bit error
(e)	$000 \rightarrow 000$	$(1-p)^3$	no errors
(f)	$000 \rightarrow 001, 010, 100$	$3(1-p)^2p$	1 bit error
(g)	$000 \rightarrow 110, 101, 011$	$3(1-p)p^2$	2 bit error
(h)	$000 \rightarrow 111$	p^3	3 bit error

can happen in three ways. In case (c) two bit errors occur, which can also hit the code word in three different ways, and so on.

The interesting question which arises deals with the amount of cases in which this coding scheme will be able to correct the bit errors. In other words, how many cases can decode the right symbol in spite of the corrupt code word bits? From Table 8.1 it can clearly be concluded that correct transmission will be achieved, that is $\hat{u} = u$ in cases (a), (b), (e) and (f). In all other cases, the majority vote decoder will make errors and will not be able to correct the bit errors in the code word. Due to the majority decoding and the fact that we only transmit the code words 111 and 000 respectively, it is clear that errors will occur if there are more than $t = 1$ bit errors in the code word, hence we can define t as

$$t = \left\lfloor \frac{n}{2} \right\rfloor \tag{8.16}$$

Where $\lfloor \rfloor$ is the 'floor' operator, that is the first integer less than or equal to the argument. The total probability of erroneous transmission can now be expressed as

$$P_{err} = \sum_{i=t+1}^{n} \binom{n}{i} p^i (1-p)^{(n-i)} = \binom{3}{2} 0.1^2 (1 - 0.1) + \binom{3}{3} 0.1^3$$

$$= 3 \cdot 0.01 \cdot 0.9 + 0.001 = 0.028 \tag{8.17}$$

where

$$\binom{n}{i} = \frac{n!}{(n-i)!\, i!}$$

is the binomial coefficient, in other words the number of ways i elements can be selected out of a set of n elements. As $n \rightarrow \infty$ in equation (8.17), $P_{err} \rightarrow 0$, in accordance with the channel coding theorem.

Intuitively from the above discussion, the greater the 'difference' between the transmitted code words, the more bit errors can be tolerated, that is before a corrupt version of one code word seems to resemble another (erroneous) code word. Assume for instance that the code uses the two code words 101 and 001 instead of 111 and 000. In this case, one single bit error is enough for the decoder to make an erroneous decision.

The 'difference' in coding context is denoted as the **Hamming distance** d. It is defined as the number of positions in which two code words differ. For example, the code words we used, 111 and 000, differ in three positions, hence the Hamming distance is $d = 3$. The two code words 101 and 001 have $d = 1$. It can be shown that a code having a minimum Hamming distance d between code words can **correct**

$$t = \left\lfloor \frac{d-1}{2} \right\rfloor \text{ bit errors in a code word} \tag{8.18}$$

Further, the same code can **detect** (but not correct)

$$y = d - 1 \text{ bit errors in a code word} \tag{8.19}$$

If there are d bit errors, we end up in another error-free code word and the error can never be detected. In some communication systems, the receiving equipment does not only output ones and zeros to the decoder. For instance, if a signal having poor strength is received, a third symbol, the **erasure** symbol, can be used. This is a 'don't care' symbol saying the signal was too weak and cannot determine if the received signal was a 1 or a 0. In this case, the decoding algorithm can fill in the most probable missing bits

$$\rho = d - 1 \text{ erasures in a code word} \tag{8.20}$$

Finally when classifying a code a triplet consisting of code word length n, number of information bits k and minimum Hamming distance d commonly stated as (n, k, d) is used. The number of parity bits is of course $n-k$. Using the minimum Hamming distance d and equation (8.18) we can figure out how many bit errors t (per code word) the code can handle. The example repetition code above is hence denoted $(3, 1, 3)$.

It is worth noting that if there are more than t bit errors in a code word, the coding will add even **more errors** and only make things worse. In such a case, a better code must be used. It might even be advantageous to not use any error-correcting code at all.

8.2.2 Linear block codes

For a general block code, we are free to select our code words arbitrarily, which for large blocks may result in extremely complicated encoders and decoders. If we demand the code to be linear, things become somewhat easier. The definition of a **linear code** is that the sum of two code words x_i and x_j must also be a code word. Addition of code words is performed component wise

$$x_i + x_j = \begin{bmatrix} x_{i1} & x_{i2} & \dots & x_{in} \end{bmatrix} + \begin{bmatrix} x_{j1} & x_{j2} & \dots & x_{jn} \end{bmatrix}$$

$$= \begin{bmatrix} x_{i1}+x_{j1} & x_{i2}+x_{j2} & \dots & x_{in}+x_{jn} \end{bmatrix}$$

$$= \begin{bmatrix} x_{m1} & x_{m2} & \dots & x_{mn} \end{bmatrix} = x_m \tag{8.21}$$

Here the addition is performed according to the algebraic rules of the number system used. For the binary case, '+' means **binary** addition, that is addition modulo 2 or exclusive OR (XOR). Thus, $0 + 0 = 0$, $1 + 0 = 0 + 1 = 1$ and $1 + 1 = 0$. For other number systems, similar rules shall apply. This algebraic topic is referred to as the Galois fields ($GF(p)$) theory. In this chapter, we will however remain faithful to the binary number system. If an interest in this particular subject is not very high, then Galois is a trivial issue.

A linear code has the nice property that the vicinity of all code words in the n-dimensional 'code space' look the same. Hence, when decoding, we can apply the same algorithm for all code words and there are many standard mathematical tools available for our help. When analysing codes, it is common to assume that the zero code word (all zeros) has been sent. The zero code word is present in all binary linear block codes, since by adding a code word to itself, using equation (8.21)

$$x_i + x_i = \begin{bmatrix} x_{i1}+x_{i1} & x_{i2}+x_{i2} & \dots & x_{in}+x_{in} \end{bmatrix} = \begin{bmatrix} 0 & 0 & \dots & 0 \end{bmatrix} = \mathbf{0} \tag{8.22}$$

the coding process in a linear block code can be readily expressed as a matrix multiplication. The information word u (dim $1 \times k$) is multiplied by the **generator matrix** G (dim $k \times n$), thus obtaining the code word x (dim $1 \times n$)

$$x_i = u_i G = u_i \begin{bmatrix} I & \vdots & P \end{bmatrix} \tag{8.23}$$

The generator matrix G defines the coding operation and can be partitioned into an identity matrix I (dim $k \times k$) and a matrix P (dim $k \times n-k$). In this way, the code will have a systematic form. The identity matrix simply 'copies' the information bits u_i into the code word, while the P matrix performs the calculations needed to determine the parity bits p_j (compare to equations (8.14)).

During the transmission, the code word x will be subject to interference, causing bit errors. These bit errors are represented by the **error vector** e (dim $1 \times n$). The received code word y (dim $1 \times n$) can be expressed as

$$y = x + e \tag{8.24}$$

(remember we are still dealing with binary addition).

Finally, when decoding a linear block code, a syndrome-based decoding algorithm is commonly used. This algorithm calculates a **syndrome vector** s (dim $1 \times k$) by multiplying the received code word y by the **parity matrix** H (dim $k \times n$). The parity matrix can be partitioned into an identity matrix I (dim $k \times k$) and the matrix Q (dim $n-k \times k$)

$$s = yH^T = y \begin{bmatrix} Q & \vdots & I \end{bmatrix}^T \tag{8.25}$$

For the binary case, if we set $Q = P^T$, the syndrome vector will be equal to the zero vector (all zero elements), that is $s = \mathbf{0}$ if there are **no errors** in the received code word that this particular code is able to detect. This can be shown by

$$S = yH^T = (uG + e)H^T = uGH^T = u[I \vdots P]\begin{bmatrix} Q^T \\ \dots \\ I \end{bmatrix}$$

$$= u(Q^T + P) = 0 \tag{8.26}$$

where of course $e = 0$ since there are no errors. If we have detectable errors, $s \neq 0$ and the syndrome vector will be used as an index in a **decoding table**. This decoding table shows the error vectors that are able to generate a specific syndrome vector. Commonly it is assumed that few bit errors in a code word is more likely than many errors. Hence, the suggested error vector $\hat{e}$ (dim $1 \times n$) having the smallest number of ones (that is bit errors), is assumed to be equal to the true error vector e. By adding the assumed error vector to the received code word including errors, a corrected code word $\hat{x}$ (dim $1 \times n$) can be obtained

$$\hat{x} = y + \hat{e} \tag{8.27}$$

From this corrected code word $\hat{x}$, only simple bit manipulations are needed to extract the received information word $\hat{u}$ (dim $1 \times k$).

Let us conclude this discussion on linear block codes with an example. Assume we have designed a linear block code LC(6,3,3), having $n = 6$, $k = 3$ and $d = 3$. The code words are in systematic form and look like

$$x = [x_1 \; x_2 \; x_3 \; x_4 \; x_5 \; x_6] = [u_1 \; u_2 \; u_3 \; p_1 \; p_2 \; p_3] \tag{8.28}$$

where the parity bits are calculated using the following expressions

$$p_1 = u_1 + u_2$$

$$p_2 = u_2 + u_3 \tag{8.29}$$

$$p_3 = u_1 + u_2 + u_3$$

Using this information, it is straightforward to write down the generator matrix for this code

$$G = [I \vdots P] = \begin{bmatrix} 1 & 0 & 0 & 1 & 0 & 1 \\ 0 & 1 & 0 & 1 & 1 & 1 \\ 0 & 0 & 1 & 0 & 1 & 1 \end{bmatrix} \tag{8.30}$$

In a similar way, the parity matrix can also be formed

$$H = [P^T \vdots I] = \begin{bmatrix} 1 & 1 & 0 & 1 & 0 & 0 \\ 0 & 1 & 1 & 0 & 1 & 0 \\ 1 & 1 & 1 & 0 & 0 & 1 \end{bmatrix} \tag{8.31}$$

Now, assume we want to transmit the information word 010, i.e. $u =$

[0 1 0]. The encoder will transmit the corresponding code word on the channel

$$x = uG = \begin{bmatrix} 0 & 1 & 0 \end{bmatrix} \begin{bmatrix} 1 & 0 & 0 & 1 & 0 & 1 \\ 0 & 1 & 0 & 1 & 1 & 1 \\ 0 & 0 & 1 & 0 & 1 & 1 \end{bmatrix} = \begin{bmatrix} 0 & 1 & 0 & 1 & 1 & 1 \end{bmatrix} \qquad (8.32)$$

This time we are lucky, the channel is in a good mode, and no errors occur, that is $e = 0$, thus

$$y = x + e = x = \begin{bmatrix} 0 & 1 & 0 & 1 & 1 & 1 \end{bmatrix} \qquad (8.33)$$

In the decoder, the syndrome vector is first calculated as

$$s = yH^T = \begin{bmatrix} 0 & 1 & 0 & 1 & 1 & 1 \end{bmatrix} \begin{bmatrix} 1 & 0 & 1 \\ 1 & 1 & 1 \\ 0 & 1 & 1 \\ 1 & 0 & 0 \\ 0 & 1 & 0 \\ 0 & 0 & 1 \end{bmatrix} = \begin{bmatrix} 0 & 0 & 0 \end{bmatrix} = 0 \qquad (8.34)$$

Since the syndrome vector is equal to the zero vector, the decoder cannot detect any errors and no error correction is needed, $\hat{x} = y$. The transmitted information is simply extracted from the received code word by

$$\hat{u} = \hat{x}\begin{bmatrix} I & \vdots & 0 \end{bmatrix} = \begin{bmatrix} 0 & 1 & 0 & 1 & 1 & 1 \end{bmatrix} \begin{bmatrix} 1 & 0 & 0 \\ 0 & 1 & 0 \\ 0 & 0 & 1 \\ 0 & 0 & 0 \\ 0 & 0 & 0 \\ 0 & 0 & 0 \end{bmatrix} = \begin{bmatrix} 0 & 1 & 0 \end{bmatrix} \qquad (8.35)$$

Now, assume we try to send the same information again over the channel, this time however, we are not as lucky as before, and a bit error occurs in u_3. This is caused by the error vector

$$e = \begin{bmatrix} 0 & 0 & 1 & 0 & 0 & 0 \end{bmatrix} \qquad (8.36)$$

Hence, the received code word is

$$y = x + e = \begin{bmatrix} 0 & 1 & 0 & 1 & 1 & 1 \end{bmatrix} + \begin{bmatrix} 0 & 0 & 1 & 0 & 0 & 0 \end{bmatrix}$$

$$= \begin{bmatrix} 0 & 1 & 1 & 1 & 1 & 1 \end{bmatrix} \qquad (8.37)$$

Table 8.2 *Decoding table for the code LC(6,3,3) in the example*

Syndrome	Possible error vectors							
000	000000	100101	010111	001011	110010	011100	101110	111001
101	100000	000101	110111	101011	010010	111100	001110	011001
111	010000	110101	000111	011011	100010	001100	111110	101001
011	001000	101101	011111	000011	111010	010100	100110	110001
100	000100	100001	010011	001111	110110	011000	101010	111101
010	000010	100111	010101	001001	110000	011110	101100	111011
001	000001	100100	010110	001010	110011	011101	101111	111000
110	101000	001101	111111	100011	011010	110100	000110	010001

The syndrome calculation gives

$$s = yH^T = \begin{bmatrix} 0 & 1 & 1 & 1 & 1 & 1 \end{bmatrix} \begin{bmatrix} 1 & 0 & 1 \\ 1 & 1 & 1 \\ 0 & 1 & 1 \\ 1 & 0 & 0 \\ 0 & 1 & 0 \\ 0 & 0 & 1 \end{bmatrix} = \begin{bmatrix} 0 & 1 & 1 \end{bmatrix} \qquad (8.38)$$

Using a decoding table (Table 8.2), we find that the eight proposed error vectors on line 4 may cause the syndrome vector obtained above. Using the maximum likelihood argument, we argue that a single bit error is more probable than a multi-bit error in our system. For this reason, we assume that the error vector in the first column to the left (containing only one bit) is the most probable error, hence

$$\hat{e} = \begin{bmatrix} 0 & 0 & 1 & 0 & 0 & 0 \end{bmatrix} \qquad (8.39)$$

The received code word is now successfully corrected (equation (8.27))

$$\hat{x} = y + \hat{e} = \begin{bmatrix} 0 & 1 & 1 & 1 & 1 & 1 \end{bmatrix} + \begin{bmatrix} 0 & 0 & 1 & 0 & 0 & 0 \end{bmatrix}$$

$$= \begin{bmatrix} 0 & 1 & 0 & 1 & 1 & 1 \end{bmatrix} = x \qquad (8.40)$$

and the transmitted information can be extracted in the same way as in equation (8.35).

The example code above has a minimum Hamming distance between code words of $d = 3$. Using equation (8.18) we find that the code is only able to correct $t = 1$, i.e. single bit errors. As pointed out earlier in this text, if there are more bit errors in a code word than t, the coding scheme will not be able to correct any errors; it may in fact aggravate the situation. This can be illustrated by the following example. If we transmit $u = [0\ 0\ 1]$, the code word will be $x = [0\ 0\ 1\ 0\ 1\ 1]$. If a **double** bit error situation occurs, $e = [0\ 1\ 0\ 1\ 0\ 0]$, we will receive $y = x + e = [0\ 1\ 1\ 1\ 1\ 1]$, that is, the same code word as in the example above. This code word will be 'corrected' and as above, we will obtain $\hat{u} = [0 \ 1 \ 0] \neq u$.

8.2.3 Cyclic codes and BCH codes

The class of **cyclic codes** is a sub-class of linear block codes. Besides the linearity property, a **cyclic shift** ('rotate') of the bit pattern should result in another code word. For example, if 111001 is a code word, then 110011 should also be a code word and so on.

Cyclic codes are often expressed using code **polynomials**. In these polynomials, a formal parameter x is used. **Note! This formal parameter x is something completely different from the code word bits x_i discussed earlier.** The formal parameter x is only a means of numbering and administrating the different bits in a cyclic code polynomial. The data bits of interest are the **coefficients** of these polynomials.

A code word $c = [c_0 \ c_1 \ \dots \ c_{n-1}]$ can hence be expressed as a code polynomial

$$c(x) = c_0 + c_1 x + c_2 x^2 + c_3 x^3 + \dots c_{n-1} x^{n-1} \tag{8.41}$$

An m step cyclic shift of an $n-1$ bit code word is accomplished by a multiplication

$$x^m c(x) \mod (x^n - 1) \tag{8.42}$$

which is performed modulo $x^n - 1$, that is after the polynomial multiplication by x^m a polynomial division by $(x^n - 1)$ should take place to keep the order of the resulting polynomial to $n-1$ or below. In bit pattern words, this operation can be explained as the multiplication performing the 'bit shifting', while the division does the 'bit rotating' (higher order bits are rotated back as low order bits).

Using this polynomial way of expressing vectors, a k bit information vector can be written as a polynomial in the formal parameter x in a similar way to equation (8.41)

$$u(x) = u_0 + u_1 x + u_2 x^2 + u_3 x^3 + \dots u_{k-1} x^{k-1} \tag{8.43}$$

where $u_0, u_1, \dots u_{k-1}$ are the bits of the corresponding information vector (but the numbering is slightly different from the vector notation discussed earlier). Still remaining with binary numbers, modulo 2 addition applies. Instead of using a generator matrix to define the code, we now use a **generator polynomial** $g(x)$, and the coding process consists of a multiplication of polynomials

$$c(x) = u(x) g(x) \tag{8.44}$$

where $c(x)$ is the code word polynomial, corresponding to the code vector being transmitted over the channel. Unfortunately, this straightforward approach does not result in code words having systematic form. This can be accomplished by first shifting the information polynomial and then dividing this product by the generator polynomial

$$\frac{x^{n-k}u(x)}{g(x)} = q(x) + \frac{r(x)}{g(x)} \qquad (8.45a)$$

where $q(x)$ is the quota and $r(x)$ the remainder. Rewriting equation (8.45a) we get

$$x^{n-k}u(x) = q(x)g(x) + r(x) = c(x) + r(x)$$

$$\Rightarrow c(x) = x^{n-k}u(x) - r(x) \qquad (8.45b)$$

The code word travels the channel as before and errors, represented by the **error polynomial** $e(x)$, may occur. The received data vector is represented by the polynomial $v(x)$

$$v(x) = c(x) + e(x) \qquad (8.46)$$

When decoding the received word expressed as $v(x)$, calculation of the **syndrome polynomial** $s(x)$, used for error detection and correction, will simply consist of division by the generator polynomial $g(x)$ or multiplication by the **parity polynomial** $h(x)$, where

$$h(x) = \frac{x^n - 1}{g(x)} \qquad (8.47)$$

The syndrome polynomial is obtained as the remainder when performing the division

$$\frac{v(x)}{g(x)} = Q(x) + \frac{s(x)}{g(x)} \qquad (8.48)$$

It is easy to show that the syndrome will depend on the error polynomial only, and if no errors are present, the syndrome polynomial will be 0. Using equations (8.46), (8.45) and (8.48) we obtain

$$v(x) = c(x) + e(x) = q(x)g(x) + e(x) = Q(x)g(x) + s(x)$$

$$\Rightarrow e(x) = (Q(x) - q(x))g(x) + s(x) \qquad (8.49)$$

This relation shows that the syndrome $s(x)$ is the remainder when dividing $e(x)$ by $g(x)$. Hence, we can calculate the syndromes for all possible error polynomials in advance. All unique syndromes can then successfully be used to identify the corresponding error polynomial $\hat{e}(x)$ and error correction of $v(x)$ can be performed as (in the binary case)

$$\hat{v}(x) = v(x) + \hat{e}(x) \qquad (8.50)$$

where $\hat{v}(x)$ is the corrected message (hopefully), from which the information $\hat{u}(x)$ can be extracted. Once again, an example will illustrate the ideas and algorithms of binary, cyclic block codes. Assume we are using a cyclic code CC(7,4,3), that is a code having $n = 7$, $k = 4$ and $d = 3$. Using equation (8.18) we find that this code can correct single bit errors, $t = 1$. Further, the code is defined by its code polynomial $g(x) = 1 + x + x^3$. In this example, we want to transmit the information 0010. This is expressed as an information polynomial

$$u(x) = x^2 \tag{8.51}$$

Multiplying this information polynomial by the generator polynomial, the code polynomial is obtained

$$c(x) = u(x)g(x) = x^2(1 + x + x^3) = x^2 + x^3 + x^5 \tag{8.52}$$

The corresponding bit sequence 0011010 is transmitted over the channel and this first time no errors occur, that is $e(x) = 0$ and

$$v(x) = c(x) + e(x) = c(x) = x^2 + x^3 + x^5 \tag{8.53}$$

At the decoder, we start by calculating the syndrome polynomial

$$\frac{v(x)}{g(x)} = \frac{x^5 + x^3 + x^2}{x^3 + x + 1} = x^2 + \frac{0}{x^3 + x + 1} = x^2 + \frac{s(x)}{x^3 + x + 1} \tag{8.54}$$

Since the remainder or the syndrome is zero, there are no errors and the received information is $\hat{u}(x) = x^2 = u(x)$. Now we try this again, but this time a single bit error occurs, having the corresponding error polynomial $e(x) = x^5$. Starting over from equation (8.53) we get (using binary addition)

$$v(x) = c(x) + e(x) = x^2 + x^3 + x^5 + x^5 = x^2 + x^3 \tag{8.55}$$

The syndrome calculation yields

$$\frac{v(x)}{g(x)} = \frac{x^3 + x^2}{x^3 + x + 1} = 1 + \frac{x^2 + x + 1}{x^3 + x + 1} = 1 + \frac{s(x)}{x^3 + x + 1} \tag{8.56}$$

Hence this time the syndrome polynomial is $s(x) = x^2 + x + 1$. Using equation (8.48) and assuming the zero code polynomial $c(x) = 0$ (OK, since the code is linear, all code words have the same 'vicinity'), we find it is easier to make a decoding table in advance

$s(x)$	$\hat{e}(x)$
0	0
1	1
x	x
x^2	x^2
$x + 1$	x^3
$x^2 + x$	x^4
$x^2 + x + 1$	x^5
$x^2 + 1$	x^6

From this table, we can see that with the given syndrome $s(x) = x^2 + x + 1$, the corresponding single bit error polynomial is $\hat{e}(x) = x^5$. Error correction yields

$$\hat{v}(x) = v(x) + \hat{e}(x) = x^2 + x^3 + x^5 \tag{8.57}$$

From this, the information $\hat{u}(x) = x^5$ can be obtained as before. In most cases, coding and decoding is not performed in the simple way outlined above, since there are smarter ways to implement the algorithms.

Shift registers and logic circuits have traditionally been used to implement cyclic code algorithms. Nowadays, these functions can readily be implemented as DSP software, using shift, rotate and Boolean computer instructions. Below, the basic ideas will be briefly presented. The underlying idea is that multiplication and division of polynomes can be viewed as **filtering operations**, and can be achieved using shift registers, in structures resembling digital FIR and IIR filters.

If we have a generator polynomial $g(x) = g_0 + g_1 x + g_2 x^2 + \ldots g_{n-k} x^{n-k}$, an information polynomial as in equation (8.43) and we want to perform the polynomial multiplication as in equation (8.44) to obtain a code vector as in equation (8.41) we get

$$\begin{aligned} c(x) &= c_0 + c_1 x + c_2 x^2 + \ldots + c_{n-1} x^{n-1} = u(x)g(x) \\ &= u_0 g_0 + (u_1 g_0 + u_0 g_1)x + (u_2 g_0 + u_1 g_1 + u_0 g_2)x^2 \ldots \end{aligned} \tag{8.58}$$

From the above expression it can be seen that the coefficients c_i can be obtained by a convolution of the coefficients of the information polynomial $u(x)$ and the coefficients of the generator polynomial $g(x)$

$$c_i = \sum_{j=0}^{i} g_j u_{i-j} \tag{8.59}$$

This operation is performed by a FIR filter (see Chapter 1), hence 'filtering' the data bit sequence u_i using a binary FIR filter with tap weights g_j will perform the polynomial multiplication, otherwise known as 'the coding operation' (equation (8.44)).

The syndrome calculation, division by $g(x)$ as in equation (8.48), can also be performed in a similar way. Consider the z-transforms of the sequences u_i and g_j, denoted $U(z)$ and $G(z)$ respectively (see Chapter 1). Since equation (8.59) is a convolution, this is equivalent to a multiplication in the z-transform domain hence the z-transform of the code polynomial coefficients $C(z)$ can easily be obtained

$$C(z) = U(z)G(z) \tag{8.60}$$

If we now require the generator polynomial to have $g_0 = 1$ and $g_{n-k} = 1$ it can be rewritten as

$$\begin{aligned} G(z) &= g_0 + g_1 z^{-1} + g_2 z^{-2} + \ldots + g_{n-k} z^{-n+k} \\ &= 1 + g_1 z^{-1} + g_2 z^{-2} + \ldots + z^{-n+k} = 1 + G'(z) \end{aligned} \tag{8.61}$$

From Chapter 1 we recall that if a FIR filter having for instance, the z-transform $G'(z)$ is put in a feedback loop, we have created an IIR filter with the transfer function

$$H(z) = \frac{1}{1 + G'(z)} = \frac{1}{G(z)} \tag{8.62}$$

This indicates that by feeding the data corresponding to $v(x)$ into a circuit consisting of a shift register in a feedback loop, we effectively achieve a polynomial division. The tap weights corresponding to $G'(z)$, are simply $g(x)$ but with $g_0 = 0$. Closer analysis shows that during the first n shifts, the quota $q(x)$ (equation (8.45a)) of the polynomial division is obtained at the output of the shift register. After these shifts, the remainder, that is the syndrome coefficients of $s(x)$ can be found in the delay line of the shift register. The syndrome can hence be read out in parallel and used as an index in a syndrome lookup table. Using information in this table, errors can be corrected. A special class of syndrome-based error-correcting decoders using the method above are known as **Meggit decoders**.

Figure 8.3 shows an encoder and channel (upper part) and decoder (lower part) for the cyclic code $g(x) = 1 + x + x^3$ in the previous example. The

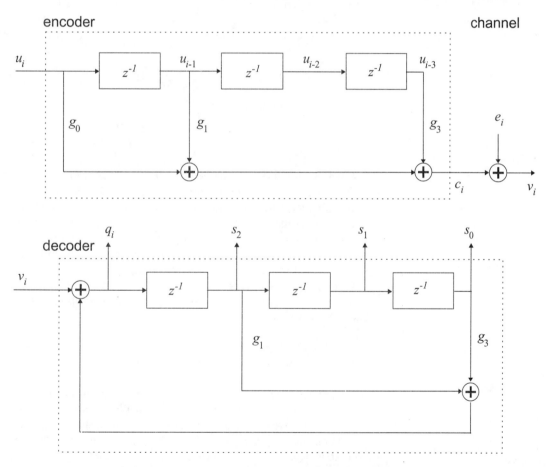

Figure 8.3 *Encoder and channel (upper) and decoder (lower) for a cyclic code having the generator polynomial* $g(x) = 1 + x + x^3$. *Encoder and syndrome calculating decoder built using shift register elements*

encoder and decoder consist of shift registers as outlined in the text above. It is an instructive exercise to mentally 'step through' the coding and syndrome calculating processes. The syndrome is obtained after $n = 7$ steps.

There are a number of standardized cyclic codes, commonly denoted **cyclic redundancy check codes (CRC)**; some of them are

CRC-4: $\qquad g(x) = 1 + x + x^4$

CRC-5: $\qquad g(x) = 1 + x^2 + x^4 + x^5$

CRC-6: $\qquad g(x) = 1 + x + x^6$

CRC-12: $\qquad g(x) = 1 + x + x^2 + x^3 + x^{11} + x^{12}$

CRC-16: $\qquad g(x) = 1 + x^2 + x^{15} + x^{16}$

CRC-CCITT: $\quad g(x) = 1 + x^5 + x^{12} + x^{16}$

A special and important class of cyclic codes are the **BCH codes,** named after Bose, Chaudhuri and Hocquenghem. This class consists of a number of effective codes with moderate block length n. The parameters of these codes are

Block length: $\qquad\qquad\qquad n = 2^m - 1$

Number of information bits: $\quad k \geqslant n - mt$

Minimum Hamming distance: $\quad d \geqslant 2t + 1$

for $m = 3, 4, 5, \ldots$

Reed–Solomon (RS) codes belong to a special type of BCH codes, working with groups of bits, in other words m-ary symbols rather than bits. The RS codes are very efficient and have the greatest possible minimum Hamming distance for a given number of check symbols and a given block length. The block length must be kept small in practice. Nevertheless, it is an important and popular type of code and is often used together with some other coding scheme, resulting in a **concatenated coding** system.

BCH and Reed–Solomon codes are for instance used in cellular radio systems, CD player systems and satellite communication systems, dating back to the 1970s.

8.2.4 Convolution codes

In the previous sections we have discussed block codes, that is codes working with a structure of fixed length, consisting of information and parity bits. **Convolution codes** work in continuous 'stream' fashion and 'insert' parity bits within the information bits according to certain rules. The convolution process is mainly a filtering process and is commonly implemented as some kind of shift register device (may be software).

A convolution encoder algorithm can be viewed as n binary FIR filters, having m step long delay lines. The information bit sequence u_l to be coded is input to the filters. Each filter j has its own binary impulse response $g_i^{(j)}$ hence the output of each filter can be expressed as a convolution sum. The output of filter j is

$$c_l^{(j)} = \sum_{i=1}^{m} g_i^{(j)} u_{l-i+1} \quad l = 1, 2, \dots L \tag{8.63}$$

where $u_1, u_2, \dots u_L$ are the information bits and $g_i^{(j)}$ the j-th **generator**, that is the impulse response of filter j. Note! The superscript is only an index, not an exponentiation operation. The code word bits $c_l^{(j)}$ will be arranged in the following way when transmitted over the channel

$$c_1^{(1)}, c_1^{(2)}, \dots c_1^{(n)}, c_2^{(1)}, c_2^{(2)}, \dots c_2^{(n)}, \dots c_{L+m}^{(1)}, c_{L+m}^{(2)}, \dots c_{L+m}^{(n)}$$

Thus, there will be $(L+m)n$ bits in the code word, and the rate of the code can be calculated by

$$R = \frac{L}{(L+m)n} \tag{8.64}$$

Further, the **constraint length** of the code, in other words the number of bits that may be affected by a given information bit is obtained from

$$n_s = (m+1)n \tag{8.65}$$

The practical algorithm (or hardware) operates as follows. Since the n delay lines will run in parallel, having the same input, only **one** delay line is needed. This line is shared by all the n filters. Initially, all the m elements in the delay line are set to zero. The first information symbol u_1 is fed to the delay line, and the outputs $c_1^{(j)}$ of all filters are calculated and sent to the output. Then the delay line is shifted one step and the next input bit u_2 enters the circuit. The filter outputs $c_2^{(j)}$ are calculated and transferred to the output as before.

This scheme is repeated until the last information bit u_L is reached. After that point, only zeros are entered into the delay line for another m steps, thus clearing the entire delay line. During this period, the **tail** of the code is generated and transmitted through the channel.

Figure 8.4 shows an example of a convolution encoder for the code $(n, k, m) = (2, 1, 2)$, with generators $g^{(1)} = (1, 0, 1)$ and $g^{(2)} = (1, 1, 1)$. Assuming an input information bit sequence u_l is $(0, 1, 0, 1)$ will result in the code word sequence $c_l^{(j)}$ (including tail) of $(00, 11, 01, 00, 01, 11)$.

The example convolution code shown here is a simple one. Convolution type codes have been used for space and satellite communications since the early 1960s. During the Voyager missions to Mars, Jupiter and Saturn during the 1970s, a couple of different convolution codes were used to ensure good data communication. Two of these codes that became NASA standards are

The **Jet Propulsion Lab** convolution code $(2, 1, 6)$, with generators

$$g^{(1)} = (1, 1, 0, 1, 1, 0, 1)$$

$$g^{(2)} = (1, 0, 0, 1, 1, 1, 1)$$

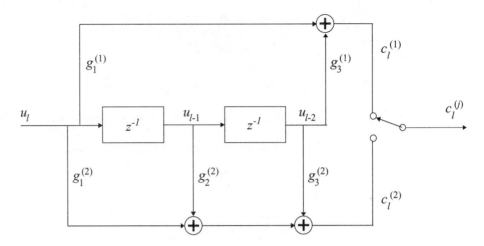

Figure 8.4 *Encoder for a simple convolution code. Parameters*: $n = 2$, $k = 1$, $m = 2$ *and generators*: $g^{(1)} = (1,0,1)$, $g^{(2)} = (1,1,1)$

The **Linkabit** (3,1,6) convolution code, with generators

$$g^{(1)} = (1, 1, 0, 1, 1, 0, 1)$$

$$g^{(2)} = (1, 0, 0, 1, 1, 1, 1)$$

$$g^{(3)} = (1, 0, 1, 0, 1, 1, 1)$$

8.2.5 Viterbi decoding

There are many ways to decode convolution-coded data. One of the most common methods is to use a **Viterbi decoder**, named after A. Viterbi. The idea of Viterbi decoding is to decode an entire sequence of information symbols and parity symbols at a time, rather than every single group (or block) of information and parity bits. Let us assume that the transmitted sequence of bits entering a channel is

$$c = (c_1, c_2, \dots c_{L+m}) = (c_1^{(1)}, c_1^{(2)}, \dots c_1^{(n)}, c_2^{(1)}, \dots c_{L+m}^{(n)})$$

as above, and $c_i = (c_i^{(1)}, c_i^{(2)}, \dots c_i^{(n)})$ is the i-th block. The received sequence, coming out of the channel is expressed in a similar way

$$v = (v_1, v_2, \dots v_{L+m}) = (v_1^{(1)}, v_1^{(2)}, \dots v_1^{(n)}, v_2^{(1)}, \dots v_{L+m}^{(n)})$$

From the discussion on channels, we remember that for a communication channel, there is a set of transition probabilities. These probabilities are denoted $P(v_i^{(j)}|c_i^{(j)})$ and are the conditional probabilities of receiving bit $v_i^{(j)}$ when transmitting $c_i^{(j)}$. If the bit error probability is independent for all bits, then the conditional probability of receiving block v_i when block c_i is transmitted is

$$P(\mathbf{v}_i|\mathbf{c}_i) = \prod_{j=1}^{n} P\left(v_i^{(j)}|c_i^{(j)}\right) \tag{8.66a}$$

In a similar way, the conditional probability of receiving the entire sequence $\mathbf{v}$, given that bit sequence $\mathbf{c}$ (including tail) is transmitted is

$$P(\mathbf{v}|\mathbf{c}) = \prod_{i=1}^{L+m} P(\mathbf{v}_i|\mathbf{c}_i) \tag{8.67a}$$

Now, the task of a maximum likelihood (ML) decoder is to maximize equation (8.67a). Taking the logarithm of equations (8.66a) and (8.67a) respectively, we obtain

$$L(\mathbf{v}_i, \mathbf{c}_i) = \log\left(P(\mathbf{v}_i|\mathbf{c}_i)\right) = \sum_{j=1}^{n} \log\left(P(v_i^{(j)}|c_i^{(j)})\right) \tag{8.66b}$$

$$L(\mathbf{v}, \mathbf{c}) = \log\left(P(\mathbf{v}|\mathbf{c})\right) = \sum_{i=1}^{L+m} \log\left(P(\mathbf{v}_i|\mathbf{c}_i)\right) \tag{8.67b}$$

Hence, given the received bit sequence $\mathbf{v}$, the ML decoder should find the bit sequence $\mathbf{c}$ that maximizes $L(\mathbf{v}, \mathbf{c})$ as in equation (8.67b). If we assume a binary symmetric channel (BSC) with error probability p, equation (8.66b) can be written

$$\begin{aligned} L(\mathbf{v}_i|\mathbf{c}_i) &= d_H(\mathbf{v}_i, \mathbf{c}_i) \log(p) + \left(n - d_H(\mathbf{v}_i, \mathbf{c}_i)\right) \log(1-p) \\ &= d_H(\mathbf{v}_i, \mathbf{c}_i) \log\left(\frac{p}{1-p}\right) + n \log(1-p) \end{aligned} \tag{8.68}$$

where $d_H(\mathbf{v}_i, \mathbf{c}_i)$ is the Hamming distance between the received binary sequence $\mathbf{v}_i$ and the transmitted sequence $\mathbf{c}_i$. The last term $n \log(1-p)$ can be neglected, since it is a constant. Further, for error probability $0 < p < 1/2$, the constant factor $\log(p/(1-p)) < 0$. Hence, we can define

$$L_i = -d_H(\mathbf{v}_i, \mathbf{c}_i) \tag{8.69}$$

and

$$S = \sum_{i=1}^{L+m} L_i = -\sum_{i=1}^{L+m} d_H(\mathbf{v}_i, \mathbf{c}_i) \tag{8.70}$$

From this we draw the conclusion that finding the maximum likelihood sequence in the case of BSC is equivalent to finding the minimum sum of Hamming distances for the blocks. This fact is utilized in the Viterbi decoding algorithm. To be able to calculate the Hamming distances at the decoder, we of course need access to the received data $\mathbf{v}_i$ and the transmitted data $\mathbf{c}_i$. Unfortunately, $\mathbf{c}_i$ is not known at the receiving site (this is why we put up the channel in the first place). Hence, using our knowledge of the operation of the encoder, we create a local copy of expected data $\hat{\mathbf{c}}_i$ for our calculations. As long as $\hat{\mathbf{c}}_i = \mathbf{c}_i$ we are doing just fine, and the correct message is obtained.

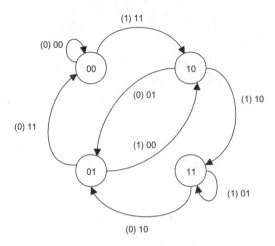

Figure 8.5 *State transition diagram for the convolution encoder shown in Figure 8.4. Digits in the nodes are the state numbers, that is the contents of the memory elements. The digits over the branches are the input information symbol u_l (in parentheses) and the output code bits $c_l^{(1)}$ and $c_l^{(2)}$.*

To demonstrate the steps needed when designing a Viterbi decoder, we will use the previous convolution code example shown in Figure 8.4. First, we draw a **state transition diagram** of the encoder. The contents of the memory elements are regarded as the states of the device. Figure 8.5 shows the state transition diagram. The digits on the branches show in parentheses the input symbol u_l and following that, the output code block $c_l = (c_l^{(1)}, c_l^{(2)})$.

Starting in state 00 (all memory elements cleared) and assuming an input information bit sequence $u = (0, 1, 0, 1)$ it is straightforward to perform the encoding process using the state transition diagram above.

input	state	output code	
0	00	00	
1	00	11	
0	10	01	
1	01	00	
0	10	01	tail
0	01	11	tail
	00		

From the above, the resulting code word sequence (including tail) is $c = (00, 11, 01, 00, 01, 11)$. Now, this state transition diagram can be drawn as a **code trellis**, a diagram showing all possible transitions (y-axis) as a function of the step number l (x-axis) (see Figure 8.6). Since we know the initial state 00, the diagram will start to fan out from one state to the left to all the possible four states. We also know that after the L steps, there will be the m steps of zeros, constituting the tail and ensuring that we end up in state 00 again. This is why we have the fan in at the right end of the trellis.

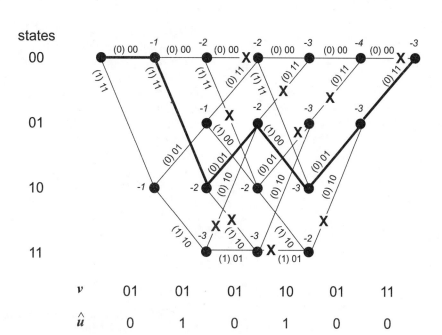

Figure 8.6 *Viterbi decoding algorithm for the example in the text, illustrated by a code trellis. States on y-axis, steps on x-axis, numbers in italics are the negative accumulated Hamming distances numbers in parentheses are the estimated information symbols, numbers on branches are the expected code blocks*

If we assume that our transmitted code sequence c suffers 3 bit errors being exposed to an error vector $e = (01, 10, 00, 10, 00)$, we receive $v = (01, 01, 01, 10, 01, 11)$. Using the trellis of Figure 8.6 and the expression (8.70), we will illustrate the Viterbi decoding algorithm.

Starting from the left, we know we are in the state 00. During the first step, the received block is $v_1 = (0, 1)$, which can be seen from the line under the trellis. Sitting in state 00 we know, from the state transition diagram in Figure 8.5, that at coding time, the encoder had two choices depending on the input information symbol. If $u_1 = 1$ the encoder could have moved to state 10 and generated the output code $c_1 = (1, 1)$, else it would have stayed in state 00 and generated $c_1 = (0, 0)$. We have received $v_1 = (0, 1)$, so when calculating L_1 using expression (8.69) we find the negative Hamming distance to be -1 in both cases. These figures are found above the respective node points in italics.

We move to the next step. Here we received $v_2 = (0, 1)$. We calculate the **accumulated** Hamming distances for all the possible transitions. For instance if we assume state 10 and transition to 11, this implies $u_2 = 1$ and $c_2 = (1, 0)$. The negative Hamming distance is -2, and since we already had -1 in state 10, it 'costs' a total -3 to reach state 11.

Starting in the third step we realize that there are two ways to reach the states. We simply proceed as before, but choose the 'cheapest' way to a node in terms of accumulated Hamming distance. The more expensive way is **crossed** out. If both ways are equally expensive, one of them is crossed and represents a cut branch. If for example we are looking for costs to reach state 01, there are two possibilities, either from state 10, cost -2 or from state 11, cost -5. We choose the former and cross out the latter.

This process is repeated until we reach the end node to the right of the trellis. If we now backtrack through the trellis, there should be only one path back to the start node to the left. While backtracking, we can easily obtain the estimated information symbols $\hat{u}$ found at the bottom of Figure 8.6. In this particular case, we managed to correct all the errors, hence $\hat{u} = u$.

This was of course, a fairly simple example, but illustrated the basics of Viterbi decoding. A similar procedure is sometimes used in equalizers to counteract intersymbol interference (ISI) in for instance radio links (e.g. cellular phones) caused by frequency selective fading (Proakis, 1989).

8.2.6 Interleaving

All the codes discussed above have the common property that only a limited number t of erroneous bits per block (or over the constraint length) can be corrected. If there are more bit errors, the coding process will most certainly make things worse, that is add even more errors.

Unfortunately, for some channels, for instance the radio channel, bit errors appear in **bursts** due to the fading phenomenon, and not randomly as assumed by the AWGN model. Hence, the probability of more than t consecutive errors to appear in a burst may be considerable. Commonly, the distance between bursts is long and the problem may be solved using for instance long block codes having large n, i.e. being able to correct a large number of bit errors. Unfortunately, the delay in the system increases with n and the complexity of most decoding algorithms increases approximately as n^3. For these reasons, long block codes are not desirable. A trick to make block codes of moderate length perform well even in burst error situations is to use **interleaving** (Ahlin and Zander, 1998).

Interleaving is a method of spreading the bit errors in a burst over a larger number of code blocks. In this way, the number of bit errors per block can be brought down to a level that can be handled by the fairly short code in use. Figure 8.7 shows the principle. We are using a block code of moderate length n, having k information bits. The incoming information symbols are encoded as before and the code words are stored row by row in the interleaving matrix. The matrix has l rows, which results in an **interleaving depth** l.

After storing l code words, the matrix is full. In Figure 8.7, $u_i^{(j)}$ is the information bit i in code word j and in a similar way $p_m^{(j)}$ is the parity bit m in code word j. The nl bits are output to the channel, but this time column by column, as $u_1^{(1)}, u_1^{(2)}, \dots u_1^{(l)}, u_2^{(1)}, u_2^{(2)}, \dots p_{n-k}^{(l)}$. At the decoder site, the reverse action takes place. The bits arriving from the channel are stored column by column, and then read out to the decoder row by row. Decoding and error correction is performed in a standard manner. What we have achieved using this scheme is that a burst of N consecutive bit errors are spread out on l code words, where every code word contains

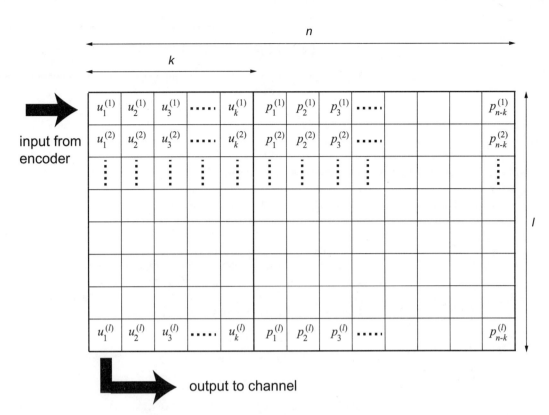

Figure 8.7 *Interleaving matrix, the encoder stores the code words row by row. The bits are then transmitted over the channel column by column. In the decoder, the reverse action takes place. Figure shows the systematic block code (n, k).*

$$t_l = \frac{N}{l} \text{ bit errors per block} \tag{8.71}$$

If $t_l \leq t$ for the code in use, the errors will be corrected. If t is the number of bits per block the code can correct, we hence can correct a burst of length

$$N = tl \text{ bits} \tag{8.72}$$

This can be viewed as: given a block code (n, k) we have created a code (nl, kl).

There are smarter ways of achieving interleaving, for instance by using **convolution interleaving** which has a shorter delay.

9 Digital signal processors

The acronym **DSP** is used for two terms, **digital signal processing** and **digital signal processor**. Digital signal processing is performing signal processing using digital techniques with the aid of digital hardware and/or some kind of computing device (signal processing can of course be analog as well). A specially designed digital computer or processor dedicated to signal processing applications is called a digital signal processor.

In this chapter, we will focus on hardware issues associated with digital signal processor chips and we will compare the characteristics of a DSP to a conventional, general-purpose microprocessor (the reader is assumed to be familiar with the structure and operation of a standard microprocessor). Furthermore, software issues and some common algorithms will be discussed.

9.1.1 Applications and requirements

Signal processing systems can be divided into many different classes, depending on the demands. One way of classifying systems is to divide them into **off-line** or **batch systems** and **on-line** or **real time systems**. In an off-line system, there is no particular demand on the data processing speed of the system, aside from the patience of the user. An example could be a data analysis system for long-term trends in thickness of the arctic ice cap. Data are collected and then stored on a data disc for instance. The data are then analysed at a relatively slow pace. In a real time system on the other hand, the available processing time is highly limited and there is a demand to follow some fast external process in time. Typical examples are digital filtering of sampled analog signals, where the filtering algorithm must be completed within the sampling period t_s. Further, in many cases no significant delay between the input signal and the output signal will be allowed. This is especially true in digital control systems, where delays may cause instability of the entire control loop (which may include heavy machinery . . .). Most applications discussed in this book belong to the class of real time systems, hence processing speed is crucial. Note: in some contexts, 'real time system' simply refers to a system being able to handle more than one piece of software at a time. There may, however, not be any requirements on processing speed at all.

Another way of classifying signal processing systems is to distinguish between **stream data** systems and **block data** systems (see Figure 9.1). In a stream data system, a continuous flow of input data is processed, resulting in a continuous flow of output data. The digital filtering system mentioned above is a typical stream data system. At every sampling instance, data are fed to the system and after the required processing time $t_p \leq t_s$ has completed, output data will be presented.

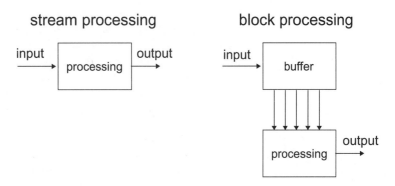

Figure 9.1 *Stream processing and block processing*

Examples of block data systems are spectrum analysers based on FFT or channel decoders. In these cases, a block of data must first be input into the system, before any computation can take place. After the processing is completed, a block of output data is obtained. A signal processing block data system often requires larger data memory than a stream data system. The most demanding applications can probably be found in the area of digital video processing. Such systems are real memory hoggers and/or require extremely high computational power. Quite often digital image processing systems are multiprocessor systems and consist of a number of processors, dedicated sub-systems and hardware. In this book, we will not consider digital image processing systems.

Another way of classifying systems relates to the numerical resolution and the dynamic range, for instance, systems that use **fixed-point** or **floating-point** arithmetic. A floating point system often makes life easier for the designer, since the need to analyse algorithms and input data in terms of numerical truncation and overflow problems does not exist as in a fixed-point design. Floating-point arithmetic may make things easier, but as pointed out in Chapter 2, it is worth noting that a 32 bit fixed-point system for instance, can have a higher resolution than a 32 bit floating-point system. Another point is that most systems dealing with real-world signals often have some analog and/or mechanical interfacing parts. These devices have a dynamic range and/or resolution of only fractions of what a floating-point signal processing system can achieve. Hence, in most practical cases, floating point systems are 'overkill'. Today, fixed-point digital processors are still less expensive and execute faster than floating-point processors.

If we try to identify the most common arithmetic operation in digital signal processing algorithms, we will find it to be the sequential calculation of a 'scalar vector product' or a convolution sum

$$y(n) = \sum_{k=a}^{b} h(k)x(n-k) = \sum_{k=a}^{b-1} h(k)x(n-k) + h(b)x(n-b) \tag{9.1}$$

Hence, the typical operation is a repeated 'multiply-add-accumulate' sequence, often denoted **MAC (multiply add accumulate)**. This is found in for instance FIR and IIR filter structures, where the input stream of samples $x(n)$ is convoluted with the impulse response coefficients $h(n)$ of the filter.

The same situation can be found in a neural network node, where the coefficients are the weights w_{ij} or in an FFT algorithm. In the latter case, the coefficients are the 'twiddle factors' W_N^{kn} (see Chapter 5). In correlators, modulators and adaptive filters, the input data sequence $x(n)$ is convoluted with another signal sequence instead of constant coefficients, else the structure is the same. Further, vector and matrix multiplication, common in block coders and decoders require the same kind of calculations.

To summarize the demand, in many digital signal processing applications we are dealing with real time systems. Hence, computational speed, that is 'number crunching' capacity is imperative. In particular, algorithms using the MAC-like operations should execute fast and numerical resolution and dynamic range need to be under control. Further, the power consumption of the hardware should preferably be low. Early DSP chips were impossible to use in battery operated equipment, for instance mobile telephones. Besides draining the batteries in no time, the dissipated power called for large heat sinks to keep the temperature within reasonable limits. On top of this, the common 'commercial' system requirements apply: the hardware must be reliable and easy to manufacture at low cost.

9.1.2 Hardware implementation

There are four different ways of implementing the required hardware:

- conventional microprocessor
- DSP chip
- bitslice or wordslice approach
- dedicated hardware, FPGA (field programmable gate array), ASIC (application specific integrated circuit).

Comparing different hardware solutions in terms of processing speed on a general level is not a trivial issue. It is not only the clock speed or instruction cycle time of a processor which determines the total processing time needed for a certain signal processing function. The bus architecture, instruction repertoire, input/output hardware, the real-time operating system and most of all, the software algorithm used will affect the processing time to a large extent. Hence, only considering **MIPS (million instructions per second)** or **MFLOPS (million floating point operations per second)** can be very misleading. When trying to compare different hardware solutions in terms of speed, this should preferably be done using the actual application. If this is not possible, benchmark tests may be a solution. The way these benchmark tests are designed and selected can of course always be subjects of discussion.

In this book, the aim is not to give exact figures of processing times, nor to promote any particular chip manufacturer ('exact' figures would be obsolete within a few years, anyhow). The goal is simply to give some approximate, **typical** figures of processing times for some implementation models. In this kind of real time system, processing time translates to the maximum sampling speed and hence the maximum bandwidth of the system.

In this text, we have used a simple straightforward 10 tap FIR filter for benchmark discussion purposes

$$y(n) = \sum_{k=0}^{9} b_k x(n-k) \tag{9.2}$$

The first alternative is a **conventional microprocessor** system, for instance an IBM-PC type system or some single-chip microcontroller board. Using such a system, development costs are minimum and numerous inexpensive system development tools are available. On the other hand, reliability, physical size and power consumption may present problems in certain applications. Another problem in such a system would be the operating system. General-purpose operating systems, for instance Windows™ (Microsoft) are not, due to their many unpredictable interrupt sources, well suited for signal processing tasks. A specialized real-time operating system should preferably be used. In many application, no explicit operating system at all may be a good solution.

Implementing the FIR filter (equation (9.2)) using a standard general purpose processor as above (no operating system overhead included) would result in a processing time of approximately $5 < t_p < 10$ µs, which translates to a maximum sampling frequency, that is $f_s = 1/t_s \leq 1/t_p$ of around 100–200 kHz. This in turn implies a 50–100 kHz bandwidth of the system (using the Nyquist criterion to its limit). The 10 tap FIR filter used for benchmarking is a very simple application. Hence, when using complicated algorithms, this kind of hardware approach is only useful for systems having quite low sampling frequencies. Typical applications could be low-frequency signal processing and systems used for temperature and/or humidity control, in other words slow control applications.

The next alternative is a **DSP chip**. DSP chips are microprocessors optimized for signal processing algorithms. They have special instructions and built-in hardware to perform the MAC operation and have architecture based on multiple buses. DSPs of today are manufactured using CMOS low voltage technology, yielding a low power consumption, well below 1 W. Some chips also have specially designed interfaces for external A/D and D/A converters. Using DSP chips requires moderate hardware design efforts. The availability of development tools is quite good, even if these tools are commonly more expensive than in the case above. Using a DSP chip, the 10 tap FIR filter (equation (9.2)) would require a processing time of approximately: $t_p \approx$ 1 µs, implying a maximum sampling frequency $f_s = 1$ MHz, or a maximum bandwidth of 500 kHz. More elaborate signal processing applications would probably use sampling frequencies around 50 kHz, a typical sampling speed of many digital audio systems today. Hence, DSP chips are common in digital audio and telecommunication applications. They are also found in more advanced digital control systems in for instance aerospace and missile control equipment.

The third alternative is using **bitslice** or **wordslice** chips. In this case, we buy sub-parts of the processor, such as multipliers, sequencers, adders, shifters, address generators etc. in chip form and design our own processor. In this way, we have full control over internal bus and memory architecture and we can define our own instruction repertoire. We hence have to do all the microcoding ourselves. Building hardware this way requires large efforts and is costly. A typical bitslice solution would execute our benchmark 10 tap FIR filter in about $t_p \approx 200$ ns. The resulting maximum sampling frequency is $f_s = 5$ MHz and the bandwidth 2.5 MHz. The speed improvement over DSP chips is not very exciting in this particular case, but the bitslice technique offers other advantages. We are for example free to select the bus width of our choice and to define special instructions for special-purpose algorithms. This type of hardware is used in systems for special-purposes, where power consumption, size and cost are not important factors.

The fourth alternative is to build our own system from gate level on silicon, using one or more **ASICs** or **FPGAs**. In this case, we can design our own adders, multipliers, sequencers and so on. We are also free mainly to use any computational structure we want. However, quite often no conventional processor model is used. The processing algorithm is simply 'hardwired' into the silicon. Hence, the resulting circuit cannot perform any other function. Building hardware in this way may be very costly and time consuming, depending on the development tools, skill of the designer and turnaround time of the silicon manufacturing and prototyping processes. Commonly, design tools are based on **VHDL (very high speed integrated circuit hardware description language)** (Sjöholm and Lindh, 1997) or the like. This simplifies the process of importing and reusing standard software-defined hardware function blocks. Further, good simulation tools are available to aid the design and verification of the chip before it is actually implemented in silicon. This kind of software tool may cut design and verification times considerably, but many tools are expensive.

Using the ASIC approach, the silicon chip including prototypes must be produced by a chip manufacturer. This is a complicated process and may take weeks or months, which increases the development time. The FPGA on the other hand is a standard silicon chip that can be programmed in minutes by the designer, using quite simple equipment. The drawback of the FPGA is that it contains fewer circuit elements (gates) than an ASIC, which limits the complexity of the signal processing algorithm. Further, FPGAs are commonly not very well suited for large volume production, due to the programming time required.

There are only two reasons for choosing the ASIC implementation method. Either we need the maximum processing speed or we need the final product to be manufactured in very large numbers. In the latter case, the development cost per manufactured unit will be lower than if standard chips were used. An ASIC, specially designed to run the 10 tap benchmark FIR filter is likely to reach a processing speed (today's technology) in the vicinity of $t_p \approx 2$ ns, yielding a sampling rate of $f_s = 500$ MHz and a bandwidth of 250 MHz. Now we are closing up to speeds required by radar and advanced video processing systems. Needless to say, when building such hardware in practise, many additional problems occur since we are dealing with fairly high frequency signals.

If yet higher processing capacity is required, it is common to connect a number of processors working in parallel in a larger system. This can be done in different ways, either in a **SIMD (single instruction multiple data)** or a **MIMD (multiple instruction multiple data)** structure. In a SIMD structure, all the processors are executing the same instruction but on different data streams. Such systems are sometimes also called vector processors. In a MIMD system, the processors may be executing different instructions. Common for all processor structures is however the demand for communication and synchronization between the processors. As the number of processors grow, the communication demands grow even faster.

Large multiprocessor systems ('super computers') of this kind are of course very expensive and rare. They are commonly used for advanced digital image processing, for solving hard optimization problems and for running large neural networks. One example of such a machine is the '**Connection Machine**' (Hillis, 1987) CM-1 (which is now replaced by CM-2). The CM-1 consists of 65536 quite simple processors, connected by a packet-switched network. The machine is fed instructions from a conventional type host computer and a specially designed computer language is used. The CM-1 has an I/O capacity of 500 Mbits/s and is capable of executing about 1000 MIPS. The machine is air cooled and dissipates about 12 kW (one tends to think of the old ENIAC, using electron tubes . . .).

An interesting thing is that this machine is only good at executing an appropriate type of algorithm, that is algorithms that can be divided into a large number of **parallel activities**. Consider our simple benchmark example, the 10 tap FIR filter. The algorithm can only be divided into 10 multiplications that can be performed simultaneously and three steps of addition (a tree of 5 groups +2 groups +1 group) which have to be performed in a sequence. Hence, we will need one instruction cycle for executing the 10 multiplications (using 10 processors) and three cycles to perform the additions, a total of four cycles. Now, if the machine consists of 65536 processors, each processor only has a capacity of $10^9/65536 = 0.015$ MIPS, which is not very impressive. If we disregard communication delays etc., we can conclude that running the 10 tap FIR filter on this super computer results in 65526 processors out of 65536 idling, and that the processing time will be in the range of 250 µs, in other words about 50 times slower than a standard PC. Our benchmark problem is obviously too simple for this machine. This also illustrates the importance of 'matching' the algorithm to the hardware architecture and that MIPS alone may not be an appropriate performance measure.

9.2 DSP versus conventional microprocessors

9.2.1 Conventional microprocessors

9.2.1.1 Architecture
A conventional microprocessor commonly uses a **von Neumann** architecture, which means that there is only one common system bus used for transfer of both instructions and data between the external memory chips and the processor (see Figure 9.2). The system bus consists of the three sub-buses: the data bus, the address bus and the control bus. In many cases, the same system bus is also used for I/O operations. In signal processing applications, this single bus is a bottleneck. Execution of the 10-tap FIR filter (equation (9.2)) will for instance require at least 60 bus cycles for instruction fetches and 40 bus cycles for data and coefficient transfers, a total of approximately

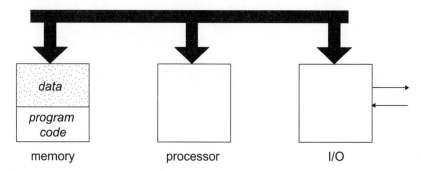

Figure 9.2 *von Neumann architecture, program code and data share memory*

100 bus cycles. Hence, even if we are using a fast processor, the speed of the bus cycle will be a limiting factor.

One way to ease this problem is the introduction of **pipe-lining** techniques, which means that an **execution unit (EU)** and a **bus unit (BU)** on the processor chip work simultaneously. While one instruction is being executed in the EU, the next instruction is fetched from memory by the BU and put into an instruction queue, feeding the instruction decoder. In this way, idle bus cycles are eliminated. If a jump instruction occurs in the program, a restart of the instruction queue however has to be performed, causing a delay.

Yet another improvement is to add a **cache memory** on the processor chip. A limited block (some thousand words) of the program code is read into the fast internal cache memory. In this way, instructions can be fetched from the internal cache memory at the same time as data is transferred over the external system bus. This approach may be very efficient in signal processing applications, since in many cases the entire program may fit in the cache and no reloading is needed.

The execution unit in a conventional microprocessor may consist of an **arithmetic logic unit (ALU)**, a **multiplier**, a **shifter**, a **floating point unit (FPU)** and some data and **flag** registers. The ALU commonly handles 2's complement arithmetics (see Chapter 2) and the FPU uses some standard IEEE floating-point formats. The **binary fractions** format, discussed later in this chapter, is often used in signal processing applications but is not supported by general-purpose microprocessors.

Besides **program counter (PC)** and **stack pointer (SP)**, the **address unit (AU)** of a conventional microprocessor may contain a number of address and **segment** registers. There may also be an ALU for calculating addresses used in more elaborate addressing modes and/or to handle virtual memory functions.

9.2.1.2 Instruction repertoire

The instruction repertoire of many general purpose microprocessors supports quite exotic addressing modes, seldom used in signal processing algorithms.

On the other hand, instructions for handling such things like **delay lines** or **circular buffers** in an efficient manner are rare. The MAC operation often requires a number of computer instructions, and loop counters have to be implemented in software, using general-purpose data registers.

Further, instructions aimed at operating systems and multi-task handling may be found among 'higher end' processors. These instructions are often of very limited interest in signal processing applications.

Most of the common processors today are of the **CISC (complex instruction set computer)** type, that is instructions may occupy more than one memory word and hence require more than one bus cycle to fetch. Further, these instructions often require more than one machine cycle to execute. In many cases a **RISC (reduced instruction set computer)** type processor may perform better in signal processing applications. In a RISC processor, no instruction occupies more than one memory word, it can be fetched in one bus cycle and executes in one machine cycle. On the other hand, many RISC instructions may be needed to perform the same function as one CISC type instruction, but in the RISC case you can get the required complexity only when needed.

9.2.1.3 Interface

Getting analog signals into and out of a general purpose microprocessor often requires a lot of external hardware. Some microcontrollers have built-in A/D and D/A converters, but in most cases these converters only have 8 or 12 bits resolution, which is not sufficient in many applications. Sometimes these converters are also quite slow. Even if there are good built-in converters, there is always the need for external sample-and-hold circuits and (analog) anti-aliasing and reconstruction filters.

Some microprocessors have built-in high-speed serial communication circuitry, **SPI (Serial peripheral interface)** or **I^2C^{TM}(Inter IC)**. In such cases we still need to have external converters, but the interface will be easier than using the traditional approach, that is, connecting the converters in parallel to the system bus. Parallel communication will of course be faster, but the circuits needed will be more complicated and we will be stealing capacity from a common single system bus.

The interrupt facilities found on many general-purpose processors are in many cases 'overkill' for signal processing systems. In this kind of real-time application, timing is crucial and **synchronous programming** is preferred. The number of **asynchronous events**, for example **interrupts** is kept to a minimum. Digital signal processing systems using more than a few interrupt sources are rare. One single interrupt source (be it timing or sample rate) or none is common.

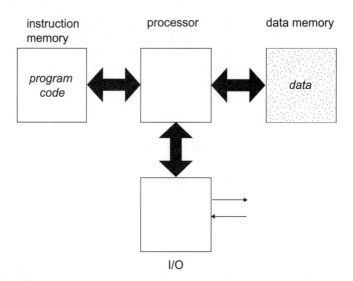

Figure 9.3 *Harvard architecture, separate buses and memories. I/O, data and instructions can be accessed simultaneously*

9.2.2 Digital signal processors

9.2.2.1 Architecture

DSP chips often have a **Harvard** architecture (see Figure 9.3) or some modified version of Harvard architecture. This type of system architecture implies that there are at least two system buses, one for instruction transfers and one for data. Quite often, three system buses can be found on DSPs, one for instructions, one for data (including I/O) and one for transferring coefficients from a separate memory area or chip.

In this way, when running an FIR filter algorithm like in equation (9.2) instructions can be fetched at the same time as data from the delay line $x(n-k)$ are fetched and as filter coefficients b_k are fetched from coefficient memory. Hence, using a DSP for the 10-tap FIR filter, only 12 bus cycles will be needed, including instruction and data transfers.

Many DSP chips also have internal memory areas that can be allocated as **data memory**, **coefficient memory** and/or **instruction memory**, or combinations of these. Pipe-lining is used in most DSP chips.

A **warning** should however be issued! Some DSP chips execute instructions in the pipe-line in a parallel, 'smart' fashion to increase speed. The result will in some cases be that instructions will **not** be executed in the same order as written in the program code. This may of course lead to strange behaviour and cumbersome troubleshooting. One way to avoid this is to insert 'dummy' instructions (for instance NOP, no operation) in the program code in the critical parts (consult the data sheet of the DSP chip to find out about pipe-line latency). This will of course increase the execution time.

The execution unit consists of at least one (often two) arithmetic logic units (ALU), a multiplier, a shifter, accumulators and data and flag registers.

The unit is designed with a high degree of **parallelism** in mind, hence all the ALUs, multipliers etc. can be run simultaneously. Further, ALUs, the multiplier and accumulators are organized so the MAC operation can be performed as efficiently as possible, with the use of a minimum amount of internal data movements. Fixed-point DSPs handle 2's complement arithmetics and binary fractions format. Floating-point DSPs use floating-point formats that can be IEEE standard (or some other non-standard format). In many cases, the ALUs can also handle both **wraparound** and **saturation** arithmetic which will be discussed later in this chapter.

Many DSPs also have ready-made **lookup tables (LUT)** in memory (ROM). These tables may be A-law and/or μ-law for companding systems and/or sine/cosine tables for FFT or modulation purposes.

Unlike conventional processors having 16, 32 or 64 bit bus widths, DSPs may have uncommon bus widths like 24, 48 or 56 bits etc. The width of the instruction bus is chosen such that a RISC-like system can be achieved, that is every instruction only occupies one memory word, and can hence be fetched in one bus cycle. The data buses are given a bus width that can handle a word of appropriate resolution at the same time as extra high bits are present to keep overflow problems under control.

The address unit is complicated since it may be expected to run three address buses in parallel. There is of course a program counter and stack pointer as in a conventional processor, but we are also likely to find a number of **index** and **pointer** registers used to generate data memory addresses. Quite often there are also one or two ALUs for calculating addresses when accessing delay lines (vectors in data memory) and coefficient tables. These pointer registers can often be **incremented** or **decremented** in a **modulo** fashion, which for instance simplifies building circular buffers. The address unit may also be able to generate the specific **bit reverse** operations used when addressing butterflies in FFT algorithms.

Further, in some DSPs, the stack is implemented as a separate **LIFO (last in first out)** register file in silicon ('hardware stack'). Using this approach, pushing and popping on the stack will be faster, and no address bus will be used.

9.2.2.2 Instruction repertoire

The special **MAC** (multiply add accumulate) instruction is almost mandatory in the instruction repertoire of a DSP. This single instruction performs one step in the summation of equation (9.2), that is it multiplies a delayed signal sample by the corresponding coefficient and adds the product to the accumulator holding the sum. Special instructions for rounding numbers are also common. On some chips even special instructions for executing the Viterbi decoding algorithms are implemented.

There are also a number of instructions that can be **executed in parallel**, to use the hardware parallelism to its full extent. Further, special prefixes or postfixes can be added to achieve **repetition** of an instruction. This is accomplished using a special **loop counter**, implemented in hardware as a special loop register. Using this register in loops, instruction fetching can be completely unnecessary in some cases.

9.2.2.3 Interface

It is common to find built-in high-speed serial communication circuitry in DSP chips. These serial ports are designed to be directly connected to codecs and/or A/D and D/A converter chips for instance. Of course, parallel I/O can also be achieved using one of the buses.

The interrupt facilities found on DSP chips are often quite simple, with a fairly small number of interrupt inputs and priority levels.

9.3 Programming DSPs

9.3.1 The development process

Implementing a signal processing function as DSP software often requires considerable design and verification efforts. If we assume that the system specification is established and the choice of suitable hardware is made, designing and implementing the actual DSP program remains.

As an example, the passband specifications for a filtering application may be determined, the filter type chosen and a transfer function $F(z)$ formulated using the z-transform. Now, starting from the transfer function, the remaining work can be described by the following checklist:

- designing the actual **algorithm**, that is the method to calculate the difference equations corresponding to the transfer function
- simulating and verifying the algorithm, using a general-purpose computer. This is often done on a non-real-time basis, using floating-point numbers and a high-level programming language, for instance C++, C or Pascal
- simulating and verifying the algorithm as above, but now using the same number format as will be used by the target DSP, for example fixed-point arithmetic
- 'translating' the algorithm computer code to the target program language applicable for the DSP, commonly C or assembly program code
- simulating the target DSP program code, using a software simulator
- verifying the function of the target DSP program code, using the 'high-level' simulations as a reference
- porting the DSP program code on the target hardware, using for example an emulator
- verifying and debugging the target system using the final hardware configuration, verifying real-time performance.

The steps above may have to be **iterated** a number of times to achieve a system having the desired performance. If unlucky, it may even turn out that the hardware configuration has to be changed to satisfy the system requirements.

When initially designing the algorithm, many alternative solutions may be evaluated to find the one that executes best using the selected hardware. Processor architecture, arithmetic performance and addressing possibilities, together with memory demands and I/O options will affect the algorithm design process.

To accomplish the 'high-level' simulations using floating point and for instance C++, C, Pascal or FORTRAN, it is also common to use some standard computational program package like MATLAB™ (MathWorks), Mathematica™ (Wolfram Research) or MathCad™ (MathSoft). General-purpose spreadsheet programs like EXCEL™ (Microsoft) have also proven to be handy in some situations. Input to the simulations may be data generated by a model or real measured and stored data.

During the 'high-level' simulations, using the same number format as the target system, the aim is to pinpoint numerical, **truncation** and **overflow** problems. As an example, when working with integer arithmetic and limited word length, the two expressions $a(b - c)$ and $ab - ac$ being equivalent from a strictly mathematical point of view, may give different results. Quite often, algorithms have to be repartitioned to overcome such problems. If the target system uses floating-point format, these simulations will in most cases be less cumbersome than in the case of fixed-point formats. Fixed-point DSPs are however more common (today), since they are less expensive and run faster than floating-point chips. Further, the numerical performance is not only a matter of number format, other arithmetic aspects like using $2's$ complement or signed integer or binary fractions etc. may also come into play. The results from these simulations will be used later in the development process as reference data.

'Translating' the algorithm into a program for the target DSP can be quite easy if the 'high-level' simulations use the same language as the target system. One such example is using the program language C, which is available for a great number of computing platforms and for many DSP chips. It is however common that some macrolanguage or pure assembly code is needed to program the target DSP. In the latter case, the program hence needs to be rewritten, commonly in assembly code. Even if high level languages such as C are becoming more common for DSPs, assembly language is the only choice if peak performance is required.

The next step is to run the DSP program code on a host computer using a software simulator. The purpose of this step is to verify the basic operation of the software.

After having debugged the DSP program code, it has to be verified and the function has to be compared to the 'high-level' simulations made earlier. 'Every bit' must be the same. If there are any deviations, the reason for this must be found.

Finally, the DSP program code is moved to the intended hardware system using an emulator, PROM (programmable read only memory) simulator or EPROM (erasable programmable read only memory). The software is now executed using the dedicated hardware.

The last step is to verify the complete system including hardware and the DSP software. In this case, the entire function is tested in real time. It should be noted that some emulators do not execute at the same speed as a real processor. There may also be timing differences regarding for instance interrupt cycles. Hence, when verifying the system in real time, using the real processor is often preferred to an emulator.

9.3.2 DSP programming languages

As mentioned above, even if there are a number of **C cross compilers** around today, is it still common to program DSPs using **assembly language**. Most human programmers are smarter than compilers. Hence, tedious assembly language programming has the potential of resulting in more compact and faster code. The development time will however be considerably longer compared to using C. For some DSPs, cross compilers for C++ and JAVA are also available. It is not easy to see that C++ or JAVA provide a better choice for signal processing software than pure C language.

The DSP assembly instruction repertoire differs somewhat from one DSP to another. Commonly, the assembly instruction can however be divided into four typical groups as detailed below.

(1) Arithmetic and logical instructions

In this group, we find instructions for adding, subtracting, multiplying and dividing numbers as well as the MAC instruction discussed earlier. There are also instructions for rounding, shifting, rotating, comparing and obtaining absolute values, etc. Many instructions may come in different versions, depending on the number format used. There may be for instance multiplication of **signed** or **unsigned** numbers, and multiplication of **integers** or **binary fractions**. In this group we can also find the standard logical, bitwise AND, OR, NOT and XOR functions.

(2) Bit manipulation instructions

This group consists of instructions for setting, resetting and testing the state of single bits in registers, memory locations and I/O ports. These instructions are handy for manipulating flags, polling external switches and controlling external indicators such as LEDs (light emitting diodes) etc.

(3) Data transfer instructions

Data transfer instructions are commonly MOVE, LOAD, STORE and so on and used to copy data to and from memory locations, registers and I/O ports. In many cases, the source and destination may have different word lengths hence care must be exercised to make sure that significant data are transferred and that sign bits etc. are properly set. Stack handling instructions also belong to this group.

(4) Loop and program control instructions

Typical instructions in this group are unconditional and conditional jump, branch and skip operations. The conditional instructions are in most cases linked to the state of the status bits in the flag register and occur in many different versions. Sub-routine jumps and returns also belong to this group. Further, we can find instructions for manipulating the hardware loop counter and for software interrupts (traps) as well as STOP, RESET and NOP (no operation).

9.3.3 The program structure

If the functions available in a proper real-time operating system are not considered, the main structure of a typical DSP program is commonly a **timed loop** or an **idling loop** and one or more **interrupt service routines**. In both cases, the purpose is to get the processing synchronized to the sampling rate. The timed loop approach can be illustrated by the following pseudo code.

```
reset:   initializing stuff
         start timer                          // sampling rate

t_loop:  do
         {
              if(timer not ready)
              {
                  background processing
              }
              else
              {                               // timed sequence
                  restart timer
                  get input
                  process
                  send result to output
              }
         } forever
```

The execution time t_p of the timed program sequence above must of course not exceed the sampling period t_s. The approach using an **idling loop and interrupt routines** assumes that interrupts are generated at sampling rate, by for instance external circuitry like A/D converters etc. This approach is shown below.

```
reset:   initializing stuff

idle:    do
         {
              background processing
         } forever

irq:                                          // timed sequence
         acknowledge interrupt
         get input
         process
         send result to output
         return from interrupt
```

In this case, the interrupt service routine must of course be executed within the sampling period t_s that is between successive interrupt signals. The latter approach is a bit more flexible and can easily be expanded using more interrupt sources and more interrupt service routines. This can be the case in

multi-rate sampled systems. One has to remember however, that the more asynchronous events like interrupt and DMA there are in a system, the harder it is to debug the system and to guarantee and verify real-time performance.

The most common DSP program has a constant inflow and outflow (stream system) of data and consists of 'simple' sequences of operations. One of the most common operations is the MAC operation discussed earlier. There are typically very few data-dependent conditional jumps. Further, in most cases only **basic addressing modes** and **simple data structures** are used to keep up execution speed.

9.3.4 Arithmetic issues

2's complement (see Chapter 2) is the most common fixed-point representation. In the digital signal processing community, we often interpret the numbers as **fractions** (fractional) rather than integers. This means that we introduce a **binary point** (not decimal point) and stay to the 'right' of the point instead of to the 'left', as in the case of integers. Hence, the weights of the binary bits representing fractions will be 2^{-1}, 2^{-2}, ...

Table 9.1 shows a comparison between fractional and integer interpretation of some binary 2's complement numbers. Fractional and integer multiplication differs by a 1 bit left shift of the result. Hence, if for instance trying fractional multiplication using a standard microprocessor or using a DSP that does not support fractions, the result must be adjusted by one step left shift, to obtain the correct result. This is because the standard multiplication rule assumes multiplication of integers. Example, multiplying 0.5 times 0.5 using the standard binary multiplication rule yields

Table 9.1 *Interpreting binary 2's complement as integer versus fraction*

2's Complement	Integer	Fraction
0111	7	.875
0110	6	.750
0101	5	.625
0100	4	.500
0011	3	.375
0010	2	.250
0001	1	.125
0000	0	.000
1111	−1	−.125
1110	−2	−.250
1101	−3	−.357
1100	−4	−.500
1011	−5	−.625
1010	−6	−.750
1001	−7	−.875
1000	−8	−1.00

```
  Binary              Fraction            Integer

    0100                0.500                4
  x 0100              x 0.500              x 4
    0000                0.250               16
    0000
    0100
    0000
    0010000 =           0.125               16

1 bit left shift yields

0100000 =               0.250
```

Most digital signal processors can handle fractional as well as integer multiplication. Some DSPs have a flag to set depending on if integer or fractional rules apply; other DSPs have separate instructions for the two kinds of multiplication. Yet other DSPs have special devices and/or instructions to perform one step shift left.

A virtue of 2's complement is that when adding a sequence of numbers whose sum you know is within bounds, all overflows and carries 'on the way' can be ignored. You will still come up with the correct result. This is very handy, since in most DSP algorithms coefficients can be scaled to guarantee that the output should be OK, thus overflows and carries can be ignored. For example

```
  Fraction            Binary              Partial sums

    .625                0101                    0101
  + .750              + 0110                    1011
  + .375              + 0011                    1110
  + .625              + 0101               (1)  0011
  − .875              + 1001                    1100
  − .625              + 1011               (1)  0111
  = .875                                   =    .875
```

We assume that the adder 'wraps around'. This will of course **not** work if the processor is using **saturating arithmetic**, that is does not wrap around. Many DSPs can support both standard (wraparound) arithmetic and saturating arithmetic, by setting flags etc. If we have very large variations in for instance an input signal and over or underflow conditions cannot be eliminated, using saturating arithmetic we will be able to preserve the sign information and have a fair chance of recovery. Saturating arithmetic basically works as a soft limiter (see Chapter 4) and can be very useful in neural network algorithms.

Multiplication of two binary numbers N bits wide, yields a result being $2N$ bits wide. This is why accumulators in DSPs are at least $2N$ bits wide, to be able to harbour results and accumulated results. At some point though, one must truncate or round back down to N bit wide numbers again.

Assume that we have an M bit 2's complement fractional number b_0, b_{-1}, b_{-2}, ... b_{-M+1}, having the weights -2^0, 2^{-1}, 2^{-2}, ... 2^{-M+1} where the left-most bit is the sign bit. If we **truncate** this word down to a width of B bits, we simply cut away all bits to the right of bit number $B-1$, i.e. the truncated word will be (assuming $M \geqslant B$): $b_0, b_{-1}, b_{-2}, \ldots b_{-B+1}$.

In this case, the error caused by the truncation operation will be

$$0 \leqslant \varepsilon < 2^{-B+1} \tag{9.3}$$

The mean error will be

$$m_\varepsilon = 2^{-B} \tag{9.4}$$

The variance of the error, or in other words the 'error noise power' will be

$$\sigma_\varepsilon^2 = \frac{2^{-2(B-1)}}{12} \tag{9.5}$$

The other possibility to reduce the word length is **rounding**. Preferably the method of 'convergent rounding' also called round-to-nearest (even) number should be used. This method works as follows:

(a) if $b_{-B}, b_{-B-1} \ldots b_{-M+1} < 1, 0, \ldots 0$ then

truncate as before

(b) if $b_{-B}, b_{-B-1} \ldots b_{-M+1} > 1, 0, \ldots 0$ then

add $b_{-B+1} = 1$ and truncate as before

(c) if $b_{-B}, b_{-B-1} \ldots b_{-M+1} = 1, 0, \ldots 0$ and $b_{-B+1} = 0$ then

truncate as before

(d) if $b_{-B}, b_{-B-1} \ldots b_{-M+1} = 1, 0, \ldots 0$ and $b_{-B+1} = 1$ then

add $b_{-B+1} = 1$ and truncate as before

Most DSPs have some kind of rounding function, but they do not all support 'convergent rounding'. The error caused by the rounding operation will be

$$-2^{-B} \leqslant \varepsilon < 2^{-B} \tag{9.6}$$

The mean error in this case will be

$$m_\varepsilon = 0 \tag{9.7}$$

The variance of the error, or in other words the 'error noise power' will be the same as for pure truncation, that is as in equation (9.5). Rounding is preferred to truncation in most cases. Truncation is easier to implement and can be performed simply using standard bit masking techniques, if a truncation program instruction is not available.

9.3.5 Data structures and addressing modes

In most cases, quite simple data structures are used in DSP software. In a typical case, there is a data vector and a coefficient vector, or two or more data vectors. In some cases, matrices are used, but since most DSP chips do not support matrix operations directly, matrices are broken down to row or column vectors and processed as a set of vectors.

To access vector structures, most DSP chips have a number of pointer registers. These registers are used for **indirect addressing**, that is the content of the register is used as an address to point to a word in the data memory. Further, the register can be **auto-incremented,** which means that the contents of the register is incremented by one, every time the register is used to point into the data memory. This can be done in two ways. Using **preincrement** the register is first incremented and then used as a pointer; if **postincrement** is used, the increment and pointing operations take place in the reverse order. There are of course also the corresponding operations for **auto-decrement**, that is decrementing the register by one.

Another addressing mode that is common is **indexed addressing**. In this case, the **effective address**, in other words the address pointed to in the data memory, is the sum of a **base register** and the **offset**, which may be the contents of an **offset register**. In this way, the base register can be used to point to the starting element of a vector or table, and the offset register is used for the actual addressing. Now the same piece of software can easily be used to process another table or vector, only the contents of the base register has to be replaced.

The addressing modes described above can be found in most conventional general-purpose microprocessors as well. In DSP chips, there are also some 'specialized' addressing modes. One such mode is auto-increment or auto-decrement using an offset other than 1. A pointer register can for instance be auto-incremented by 5 or some other constant stored in a register.

Quite often, one or more of the data vectors in a DSP program are also used as a **delay line** in for instance FIR filters. This means that a vector in data memory of length M data words is used to store the present value of our sampled signal $x(n)$ and the $M-1$ 'old' sampled values $x(n-1)$, $x(n-2)$, ... $x(n-M+1)$:

$0000+M-1$	$x(n-M+1)$
$\vdots$	$\vdots$
$\vdots$	$\vdots$
$\vdots$	$\vdots$
0002	$x(n-2)$
0001	$x(n-1)$
0000	$x(n)$

where we have assumed that the delay line vector starts at address 0000, in other words the base address is 0000. Now, for every time the sampling clock ticks, a new sampled value will arrive, and the entire delay line must be updated. This can be done in two ways. Either we are using a straightforward **static list** method or a somewhat more complicated **dynamical list** method. The static list method is based on moving the contents of all the memory cells in the vector one step 'upwards', except the oldest element at address $0000+M+1$ that will be dropped. The new element will then be stored at the bottom of the vector in address 0000. This method is easy to implement and understand. The drawback is however, that for every new sample, we have to make $M-1$ 'extra' data moves, thus consuming time.

A smarter approach is the dynamical list method. In this case, a **circular buffer** and a **start-of-list pointer** are used. The logical start of the vector is pointed to by the start-of-list pointer and can be any address in the memory block, not only address 0000 as above. As soon as a new sample arrives, we only need to move the start-of-list pointer 'downwards' one step. At this address, the oldest sample can be found, since we are now dealing with a circular list. The new sample is stored at this position, thus overwriting the oldest sample. In this way, no shuffling of data is needed like in the static list method. Only the start-of-list-pointer needs to be decremented. To handle a circular buffer, pointers need to 'wrap around' when reaching the top or bottom of the memory area allocated for the vector. This calls for **modular addressing**, supported by many DSPs. General-purpose processors do not support this type of addressing, hence extra program instructions are required to test the pointers and 'wrap around' when needed. If the size of the buffer is 2^k, where k is a positive integer, modular addressing can be achieved by simply ANDing the address pointer with an appropriate bit mask. Many DSP chip have specialized addressing hardware, supporting almost any buffer length, for example 17 or 38 etc.

Dealing with for instance some FFT algorithms, using 'butterfly' type computing strategies, **bit reverse addressing**, supported by many DSPs is handy. The DSP (as most other digital computers) performs the calculations sequentially. Equations (5.7) and (5.8) in Chapter 5 describe a 4 point FFT. It is very simple, but can serve as an example to demonstrate the ideas behind bit reverse addressing. Looking at equation (5.7), describing the computational 'butterflies', we find that we need to access the input elements $x(0)$, $x(1)$ . . . etc. The smartest way (in terms of calculation order) to access these elements is not in the 'normal' sequence 0, 1, 2, 3 but in the bit

Table 9.2 *'Normal' order addressing versus bit reverse order addressing*

Normal order		Bit reverse order	
0	000	000	0
1	001	100	4
2	010	010	2
3	011	110	6
4	100	001	1
5	101	101	5
6	110	011	3
7	111	111	7

reversed order 0, 2, 1, 3. If the transform has more than four input values, the gain in calculation time will of course be greater. So bit reverse addressing is a way of calculating the pointer values so that an FFT algorithm accesses the elements of the input vector in the 'smartest' order. The bit reverse addressing is obtained by 'mirroring' the address word, so the most significant bit becomes the least significant bit and vice versa. Table 9.2 shows a 3 bit normal and bit reverse addressing scheme.

9.4 Implementation examples

9.4.1 FIR filter

We will use a simple third-order low-pass FIR filter (see Chapter 1) as an example. The filter specification is 0 dB attenuation in the passband $0 < q < 0.2$ and an attenuation of at least 20 dB in the stop band $0.4 < q < 0.5$, where the frequencies are given in 'fnosq', that is relative frequency (see Chapter 1, equation (1.8)). Using a standard filter design program packet, the resulting filter has the transfer function.

$$H(z) = b_0 + b_1 z^{-1} + b_2 z^{-2} + b_3 z^{-3} \qquad (9.8)$$

where the filter coefficients are

$b_0 = 1.000000$

$b_1 = 0.2763932$

$b_2 = 0.6381966$

$b_3 = b_1 = 0.2763932$

The corresponding difference equation, that is the filter algorithm can easily be obtained from the transfer function (9.8)

$$y(n) = b_0 x(n) + b_1 x(n-1) + b_2 x(n-2) + b_3(n-3) \qquad (9.9)$$

The first thing to investigate is the risk of overflow during some steps in the calculation. Fortunately, this task is very simple for FIR filters. We can easily calculate the maximum gain of the filter. Since it is a straightforward low-pass filter, maximum gain will be found at frequency $q = 0$, or in other words 'DC'.

If we assume a constant DC input, that is $x(n) = x(n-1) = x(n-2) = 1$, the gain of the filter is simply the sum of the coefficients

$$G(0) = \sum_{i=0}^{3} b_i$$

$$= 1.000000 + 0.2763932 + 0.6381966 + 0.2763932 = 2.2 \qquad (9.10)$$

If we for instance assume that the maximum word length of the input signal is 24 bits, and the coefficients are 24 bits, this yields 48 bits out from the multiplier. Further, if we assume that the accumulators are 56 bits wide, we have a margin of 8 bits corresponding to a multiplication (that is gain) of

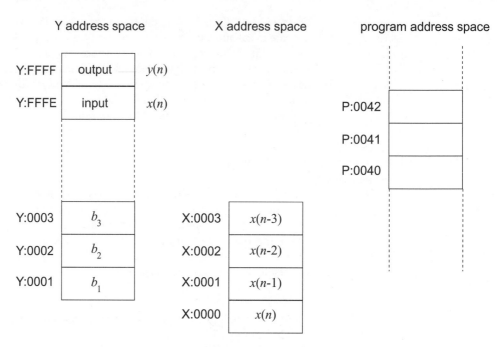

Figure 9.4 *Memory usage for the static list, third-order FIR filter example*

256 times. Since the maximum gain of the filter is roughly 2, we will never use more than 49 bits out of the 56 bits in the accumulators, hence no overflow problem will occur. If we had suspected overflow problems, we could have **scaled** all the filter coefficients by a proper scaling constant, to prevent overflow in the accumulator.

Now, our first algorithm will use a simple straightforward **static list** approach. The memory usage can be seen in Figure 9.4. There are three address spaces, the Y data memory address space, the X data memory address space and the P program memory address space. The corresponding letters are put in front of the hexadecimal addresses for clarity. We are using memory mapped I/O and the ports are mapped into the Y address space, where also the coefficient memory is located. The input port has address Y:FFFE and the output port Y:FFFF. The filter coefficients b_1, b_2, b_3 are stored at locations Y:0001, Y:0002 and Y:0003 respectively. Since $b_0 = 1$, it has been omitted.

The delay line is mapped in the X data memory according to Figure 9.4 and the program code is of course stored in the program memory, mapped into P address space. The program starts at address P:0040. The algorithm can be described by the following pseudo code.

```
init:   reset pointers
        load filter coefficients into Y memory
        clear delay line in X memory
```

```
loop:       do
            {
                get input data into delay line
                restart pointers
                clear accumulator A
                for 3 loops
                {
                    get x value from delay line
                    point to next x value
                    get coefficient
                    point to next coefficient
                    multiply and accumulate in A
                    move x element one step in delay line
                }
                add x(n) since this coefficient is 1
                round to 24 bits
                send to output
            } forever
```

As an example, assume we make an assembly program for the Motorola DSP 56001 digital signal processor chip ('DSP56000/56001 Digital Signal Processor User's Manual', 1989). The R0 register will be used as pointer to X memory. The R1 register will point to Y memory and the offset register N0 will hold the fixed 1 step offset used when moving elements in the delay line one step. The first part of the algorithm is only initializations and will only be run once, after a reset. The actual filtering action takes place in the filter loop. This loop would normally be timed in some way, to be synchronized to the sampling rate. This detail has been omitted in this example, and we are running the loop at maximum speed. The actual assembly code for the program is shown below.

```
        ; initializing

        ; init pointers
P:0040  init:       MOVE    #$3,R0              ; pointer for X data
P:0041              MOVE    #$1,R1              ; pointer for Y data
P:0042              MOVE    #$1,N0              ; offset, X pointer

        ; load filter coefficients into Y memory
P:0043              MOVE    #.2763932,X1        ; load b1
P:0045              MOVE    X1,Y:(R1)+          ; to Y(1), increment pointer
P:0046              MOVE    #.6381966,X1        ; load b2
P:0048              MOVE    X1,Y:(R1)+          ; to Y(2), increment pointer
P:0049              MOVE    #.2763932,X1        ; load b3
P:004B              MOVE    X1,Y:(R1)           ; to Y(3)

        ; clear delay line vector in X memory
P:004C              MOVE    #$0,X1
P:004D              MOVE    X1,X:(R0)-          ; X(n-3)=0, decrement pointer
P:004E              MOVE    X1,X:(R0)-          ; X(n-2)=0, decrement pointer
P:004F              MOVE    X1,X:(R0)           ; X(n-1)=0
```

```
          ; the actual filter loop
          ; get input data
P:0050    floop:    MOVE    Y:$FFFE,X1        ; from memory mapped port
P:0052              MOVE    X1,X:$0000        ; to X(n)
          ; restart pointers
P:0053              MOVE    #$3,R0            ; pointer to X mem, delay line
P:0054              MOVE    #$3,R1            ; pointer to Y mem, coefficients
          ; clear accumulator
P:0055              CLR     A
          ; the convolution process
P:0056    DO                #$3,$005D         ; hw loop 3 times, exit to $005D
P:0058              MOVE    X:(R0)-,X0        ; get data, decrement pointer
P:0059              MOVE    Y:(R1)-,Y0        ; get coeff, decrement pointer
P:005A              MAC     X0,Y0,A           ; the MAC operation
          ; the data shuffling in the delay line
P:005B              MOVE    X:(R0),X0         ; get element
P:005C              MOVE    X0,X:(R0+N0)      ; move one step 'upwards'
          END                                 ; end of loop
P:005D              ADD     X0,A              ; add x(n) since b₀ = 1
          ; round result back to 24 bits and send to output
P:005E              RND     A                 ; convergent rounding
P:005F              MOVE    A1,Y:$FFFF        ; result in A1 to output
P:0060              JMP     $0050             ; get next sample
```

The first part of this software is the initialization of pointers and memory, which requires 16 words of program memory. The # character denotes an immediate constant and the expression MOVE X1,Y:(R1)+ means copy data from register X1 to Y memory at the address pointed to by R1 and postincrement of R1. This initialization part requires 36 machine cycles to execute. Using a 20 MHz clock, the cycle time is 100 ns, hence the initialization takes 3.6 μs. Maybe it is possible to find a smarter way of writing this piece of software, but it is not worth the effort since this code is only executed once, following a reset. It is better to concentrate on making the actual filter loop faster if possible, because the execution time of this loop is the upper limit sampling rate of the system.

In the filter loop, input data is first retrieved from the input port and sent to $x(n)$ in the delay line in X memory. The pointers for X memory (delay line) and Y memory (coefficients) are restarted to point to the oldest element $x(n-3)$ and the coefficient b_3 respectively. Since we will use accumulator A in the summation process, the accumulator is cleared.

The inner loop performing the convolution in a sequential fashion is implemented as a hardware loop, meaning that we use the internal hardware loop counter on the chip. This loop starts with DO and ends with END. After DO follows the number of repetitions and the exit address, where execution continues after the loop is finished. Inside the inner loop, first an x value from the delay line is fetched as well as the corresponding filter coefficient. The heart of the algorithm is the MAC operation, that is multiplying the x value (in register X0) by the filter coefficient (in register Y0) and adding this product to the contents of accumulator A.

Table 9.3 *The sequential calculation of the difference equation (every line is an iteration of the inner loop)*

X:0000	X:0001	X:0002	X:0003	A
$x(n)$	$x(n-1)$	$x(n-2)$	$x(n-3)$	0
$x(n)$	$x(n-1)$	$x(n-2)$	$x(n-2)$	$x(n-3)b_3$
$x(n)$	$x(n-1)$	$x(n-1)$	$x(n-2)$	$x(n-3)b_3 + x(n-2)b_2$
$x(n)$	$x(n)$	$x(n-1)$	$x(n-2)$	$x(n-3)b_3 + x(n-2)b_2 + x(n-1)b_1$
$x(n)$	$x(n)$	$x(n-1)$	$x(n-2)$	$x(n-3)b_3 + x(n-2)b_2 + x(n-1)b_1 + x(n)b_0$

After the MAC operation follows the data shuffling needed to update the delay line. The present *x* value is moved 'upwards' in the list, thus 'ageing' one sample time. After the inner loop has been processed three times, the calculation of the difference function (9.9) is almost finished. Only the term $x(n)b_0$ is missing. Since $b_0 = 1$, we simply add $x(n)$ to the accumulator A. The result is then rounded back from 56 bits to 24 bits using convergent rounding and then finally sent to the memory-mapped output parallel port. The filter loop is now ready to fetch the next input sample. The sequential calculation of function (9.9) is shown in Table 9.3.

The filter loop code requires 18 words of program memory and executes in 96 machine cycles, corresponding to 9.6 µs at 20 MHz clock speed. This means that the maximum sampling rate using this software is 104 kHz. It is important to note that we have not used the possibilities of parallel data moves inherent in the chip. The DSP 56001 can do better, which we will demonstrate in the next example.

We will now show a more sophisticated way of implementing the same filter (equation (9.9)) as above, using a **dynamic list** approach and a **modular programming** style. The dynamic list approach implies that we do not need to shuffle the delay line data around. Instead, we implement the delay line as a circular buffer, using modulo addressing and with use of a start-of-list pointer. Thus, by moving the start-of-list pointer one step, the entire delay line 'ages' in a jiffy. The memory usage is shown in Figure 9.5. The main difference compared to the previous example is that we have added the coefficient b_0 to the Y memory vector to make the program more general. We cannot expect $b_0 = 1$ in the general case. Further, since we are now using a delay line in the form of a circular buffer, the position of the *x* values having different delays will vary, for example there is no fixed place for $x(n-2)$ in the buffer. It will change continuously according to the start-of-list pointer.

Another improvement is a more structured use of program memory. Starting from address P:0040 there is a block of constants, followed by an area of sub-routine code from P:0050 and at address P:0070, the main program code can be found. Hence, this address is also the entry point of the system. The constant area contains the filter coefficients. In this example we are not using immediate type program instructions to load coefficients. In many cases, it is advantageous not to 'embed' constants in the code. It is better to collect

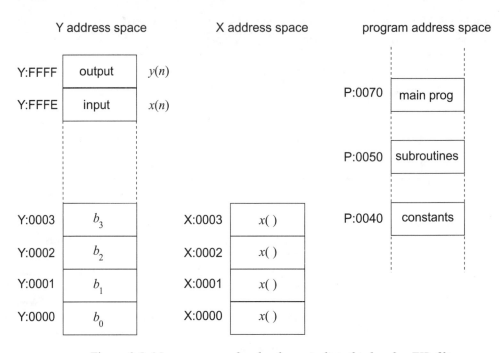

Figure 9.5 *Memory usage for the dynamic list, third order FIR filter example*

all constants in an easily found block in program memory. This is especially true if we need to change the constants frequently and/or if the constants are used in many different places of the program. The algorithm can be described by the following pseudo code

```
init:    reset pointers
         load coefficients into Y memory from program memory
         clear delay line in X memory
         return

filter:  get input data into delay line
         clear accumulator A
         for 4 loops
         {
            get x value from delay line
            point to next x value
            get coefficient
            point to next coefficient
            multiply and accumulate in A
            move x element one step in delay line
         }
         round to 24 bits
         send to output
         return
```

```
                main:   init
                        do
                        {
                            filter
                        } forever
```

In the assembly code, register R4 will be used to point to coefficients in Y memory and R0 will point to x values in the delay line vector in X memory. Since we will now be working with circular buffers, modulo addressing will be needed. For this reason, the modulo registers M4 and M0 will be loaded with 4. Further, since we always go through the entire delay line from oldest to newest x value once every sample period, we do not need an explicit start-of-list pointer. We automatically know where the list starts.

Further, in this example we will use the parallel execution feature, which makes the entire 'inner' loop only **one** instruction long. For this reason, we will use the REP (repeat) function instead of the DO END hardware loop. The assembly program is shown below.

```
            ; coefficients
P:0040                  DC      .9999999        ; b0
P:0041                  DC      .2763932        ; b1
P:0042                  DC      .6381966        ; b2
P:0043                  DC      .2763932        ; b3

            ; initializing
            ; init pointers and modulo registers
P:0050      init:       MOVE    #$0043,R1       ; temporary pointer
P:0051                  MOVE    #$0003,R4       ; pointer for Y memory
P:0052                  MOVE    #$0000,R0       ; pointer for X memory
P:0053                  MOVE    #$0003,M0       ; modulo 4, circ addressing X
P:0054                  MOVE    M0,M4           ; modulo 4, circ addressing Y
            ; load coefficients to Y and clear delay line in X
P:0055                  MOVE    #$0,X0          ; set X0=0
P:0056          DO      #$4,$005A               ; hw loop 4 times, exit to $005A
P:0058                  MOVE    P:(R1)-,Y0      ; get coeff from program memory
P:0059                  MOVE    X0,X:(R0)+ Y0,Y:(R4)-
                                                ; clear X, increment pointer
                                                ; store coeff, decrement pointer
                END
P:005A                  RTS                     ; return from subroutine

            ; the filter
P:0060      filter:     MOVEP   Y:FFFE,X:(R0)-  ; get input, store in delay line
            ; clear accumulator get x value to X0 and coeff to Y0
P:0061                  CLR     A X:(R0)-,X0 Y:(R4)-,Y0
P:0062          REP     #$3                     ; repeat next instr 3 times
P:0063                  MAC     X0,Y0,A X:(R0)-,X0 Y:(R4)-,Y0
                                                ; MAC operation
                                                ; get next x value
                                                ; get next coeff
```

```
P:0046                    MACR   X0,Y0,A              ; MAC operation and rounding
                 ; send to output
P:0047                    MOVEP  A1,Y:$FFFF           ; to output
P:0048                    RTS                         ; return from subroutine

                 ; main program
P:0070           main:    JSR    $0050                ; run init
P:0071                    JSR    $0060                ; run filter
P:0072                    JMP    $0071                ; forever
```

This software makes use of the possibilities of **parallel execution**; up to three instructions can be executed simultaneously under certain circumstances. These instructions are written on the same line. For instance MOVE X0,X:(R0)+ Y0,Y:(R4)- means copy contents of X0 to X memory address pointed to by R0 and postincrement R0 and copy contents of Y0 to Y memory address pointed to by R4 and postdecrement R4.

Another example is CLR A X:(R0)-,X0 Y:(R4)-,Y0 which means clear accumulator A and copy contents of X memory address pointed to by R0 to X0 and postdecrement R0 and copy contents of Y memory address pointed to by R4 to Y0 and postdecrement R4.

Yet another (the best one of all) MAC X0,Y0,A X:(R0)-,X0 Y:(R4)-,Y0 which means multiply contents of X0 (x value) by contents of Y0 (coefficient) and add to contents of accumulator A, store in A and copy contents of X memory address pointed to by R0 to X0 and postdecrement R0 and copy contents of Y memory address pointed to by R4 to Y0 and postdecrement R4.

Making a table of the same type as Table 9.3 for this example is left as an exercise for the reader. The program occupies a total of 25 words or program memory of which 4 words are constants, 11 words are used by the init routine, 7 by the filter routine and 3 by the main program. The init function executes in 62 machine cycles, that is 6.2 μs and the filter in 64 cycles, or in other words 6.4 μs. This means that the maximum sampling rate is 156 kHz.

9.4.2 IIR filter

In this example a second-order low-pass IIR filter (see Chapter 1) will be used. The filter specification is: 0 dB attenuation in the pass band $0 < q < 0.1$ and an attenuation of at least 30 dB in the stop band $0.4 < q < 0.5$, where the frequencies are given in 'fnosq', that is relative frequency (see Chapter 1, equation (1.8)). Using a standard filter design program packet, the resulting filter has the transfer function

$$H(z) = \frac{b_0 + b_1 z^{-1} + b_2 z^{-2}}{1 - a_1 z^{-1} - a_2 z^{-2}} \tag{9.11}$$

where the filter coefficients are

$$b_0 = 1.000000$$

$$b_1 = 1.79941$$

$$b_2 = 1.000000$$

$$a_1 = -0.299624$$

$$a_2 = 0.195921$$

This is a 'combined' IIR and FIR filter (see Chapter 1, Figure 1.7) and the corresponding difference equation is

$$y(n) = b_0 x(n) + b_1 x(n-1) + b_2 x(n-2) + a_1 y(n-1) + a_2 y(n-2) \quad (9.12)$$

A standard method is to divide this expression into two equations representing the FIR and IIR portions to simplify the implementation. Starting out from equation (9.11) we can separate the IIR and FIR parts

$$H(z) = \frac{Y(z)}{X(z)} = \frac{E(z)}{X(z)}\frac{Y(z)}{E(z)} = F(z)\,G(z) = \frac{b_0 + b_1 z^{-1} + b_2 z^{-2}}{1 - a_1 z^{-1} - a_2 z^{-2}}$$

$$= \frac{1}{1 - a_1 z^{-1} - a_2 z^{-2}}(b_0 + b_1 z^{-1} + b_2 z^{-2}) \quad (9.13)$$

The IIR part $F(z)$ has the difference equation

$$e(n) = x(n) + a_1 e(n-1) + a_2 e(n-2) \quad (9.14)$$

and the FIR part $G(z)$

$$y(n) = b_0 e(n) + b_1 e(n-1) + b_2 e(n-2) \quad (9.15)$$

Hence, we do not put $x(n)$, $x(n-1)$, ... in the delay line, it is smarter to put the intermediate signal $e(n)$ into the delay line, that is $e(n)$, $e(n-1)$, ..., where $e(n)$ is the output from the IIR part of the filter (see also Figure 1.7, Chapter 1).

Since we are now dealing with an IIR filter having poles in the transfer function, we must make sure the filter is stable, in other words the poles are within the unit circle in the complex z-plane. There are many standard methods to check stability, but even if the filter is stable, it may have an oscillatory impulse response and some **resonant peak** with a very high gain. Two problems arise, firstly, the high gain may amplify weak roundoff and truncation noise in the filter to considerable output levels ('phantom' output) and secondly, overflow problems are likely to appear. In many cases it is harder to determine good scaling factors for IIR filters than for FIR filters, where the magnitude of the signals in the filter are easier to calculate. For these reasons, IIR structures often have low orders typically 2 or 3. If a higher-order IIR filter is needed, a number of **cascaded** second order filters are used. The same program code can be used (sub-routine) for all the cascaded filters, only separate delay lines and coefficient vectors are needed. Note, if considering the roundoff noise, reordering the cascaded IIR filters may change the noise level.

Simulating the algorithm is probably the best way to pinpoint noise and overflow problems in IIR type filters. Running such a simulation of equations (9.14) and (9.15) the maximum expected gain is found to be about 3.5 times. The filter is well damped, and no significant 'ringing' can be seen. Hence, no problems are expected inside the filter itself. There are however external sub-systems to consider as well. If our input signals originate from an A/D converter and the output is connected to a D/A converter, this 'extra' gain may cause problems. To avoid overflow in the D/A converter, we decide to reduce the maximum gain of the filter by a factor of 2. This is done by scaling the filter coefficients b_0, b_1, b_2 of the FIR part with a scaling factor 0.5. The scaled filter coefficients are

$$0.5b_0 = 0.500000$$

$$0.5b_1 = 0.89970$$

$$0.5b_2 = 0.500000$$

$$a_1 = -0.299624$$

$$a_2 = 0.195021$$

In this last example, a non-standard 'smart' type algorithm using a circular buffer approach for the coefficients and two registers X0 and X1 for the delay line will be described. This algorithm is not very 'neat' from the programming point of view, but it is **quick** and **compact**. The memory usage can be seen in Figure 9.6. The two accumulators A and B are used to store intermediate results. The assembly code for the algorithm is shown below.

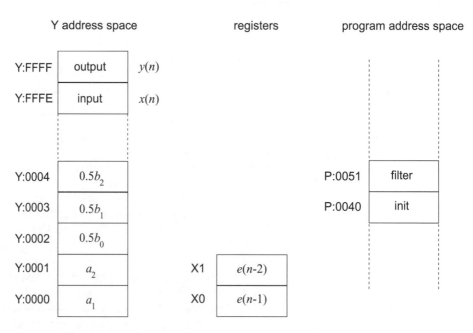

Figure 9.6 *Memory usage for the second order IIR filter example*

```
                 ; initializing
                 ; init pointer for coefficients
P:0040   init:    MOVE    #$0,R4              ; pointer for Y memory
P:0041            MOVE    #$4,M4              ; modulo 5 addressing in Y mem
                 ; load coefficients into Y memory
P:0042            MOVE    #-.299624,X0        ; load a1
P:0044            MOVE    X0,Y:(R4)+
P:0045            MOVE    #.195021,X0         ; load a2
P:0047            MOVE    X0,Y:(R4)+
P:0048            MOVE    #.500000,X0         ; load 0.5b0
P:004A            MOVE    X0,Y:(R4)+
P:004B            MOVE    #.29970,X1          ; load 0.5b1
P:004D            MOVE    X1,Y:(R4)+
P:004E            MOVE    X0,Y:(R4)+          ; load 0.5b2 = 0.5b0
                 ; clear delay line
P:004F            MOVE    #$0,X0              ; e(n-1) = 0
P:0050            MOVE    X0,X1               ; e(n-2) = 0
                 ; the filter
P:0051   filter:  CLR     A                   ; clear accumulator A
P:0052            CLR     B Y:(R4)+,Y0        ; clear accumulator B
                                             ; get a1 to register Y0
P:0053            MOVEP   Y:$FFFE,A1          ; get input to accumulator A
P:0054            MAC     X0,Y0,A Y:(R4)+,Y0  ; x(n)+a1e(n-1) to A
                                             ; get a2 to register Y0
P:0055            MACR    X1,Y0,A Y:(R4)+,Y0  ; x(n)+a1e(n-1)+ a2e(n-2) to A
                                             ; round to 24 bits
                                             ; get 0.5b0 to register Y0
P:0056            MOVE    A1,Y1               ; e(n) to register Y1
P:0057            MAC     Y0,Y1,B Y:(R4)+,Y0  ; 0.5b0e(n) to B
                                             ; get 0.5b1 to register Y0
P:0058            MAC     X0,Y0,B Y:(R4)+,Y0  ; 0.5b0e(n)+0.5b1e(n-1) to B
                                             ; get 0.5b2 to register Y0
P:0059            MACR    X1,Y0,B X0,X1       ; 0.5b0e(n)+0.5b1e(n-1)+
                                             ; +0.5b2e(n-2)to B
                                             ; e(n-1) to X1 (time passes by)
P:0060            MOVE    A1,X0               ; e(n) to X0
P:0061            MOVEP   B1,Y:$FFFF          ; output
P:0062            JMP     $0051               ; next sample
```

The program occupies 29 words of program memory, of which 17 words are for the initialization procedure and 12 for the filter. Initialization executes in 3.6 μs and the filtering function in 6.2 μs assuming 20 MHz clock speed. Hence 160 kHz sampling rate is maximum.

9.5 Where to?

The above simplified program code segments are only intended as examples and the DSP chip used is one out of many. It is not very risky to predict that future DSP chips will be faster, more complex and cheaper. In some respects, there will probably be a merging of conventional microprocessor chip technology, DSPs and gate-array structures. There are for instance general-purpose microprocessor chips around today which have MAC instructions and other typical DSP features. New, improved simulators, compilers and other development tools will also be available making life easier for the designer.

DSP chips are used in many embedded systems today in large-volume consumer products like cellular mobile telephones. As the processing speed of the DSPs increases, new applications will develop continuously. One interesting area is **radio technology**. An average cellular mobile telephone today contains a number of DSP chips. Classical radio electronics circuitry occupies only a small fraction of the total printed circuit board area. As DSP chips get faster, more and more of the radio electronics will disappear and typical radio functions like filtering, mixing, oscillating, modulation and demodulation will be implemented as DSP software rather than hardware. This is true for radios in general, not only cellular mobile telephones. For instance radio systems based on **WCDMA (wideband code division multiple access)** and **UWB (ultra wideband)** will depend heavily on DSP technology. There will be an era of s**oftware defined radio (SDR)**.

DSPs will improve but still the **programming** of DSPs will be more and more complex and demanding. Not only good programming skills are required, but also considerable knowledge in signal theory, numerical methods, algorithm design and mathematics.

References

(1)   Ahlin, L., Zander, J. (1998). *Principles of Wireless Communications*. Student-litteratur. ISBN 91–44–00762–0

(2)   Anderson, B.D.O., Moore, J.B. (1979). *Optimal Filtering*. Prentice-Hall. ISBN 0–13–638122–7.

(3)   Åström, Karl J., Wittenmark, Björn (1984). *Computer Controlled Systems, Theory and Design*. Prentice-Hall. ISBN 0–13–164319–3.

(4)   Bergh, J., Ekstedt, F., Lindberg, M. (1999). *Wavelets*. Studentlitteratur. ISBN 91–44–00938–0.

(5)   Burrus, C.S., Parks T.W. (1985). *DFT/FFT and Convolution Algorithms*. Wiley. ISBN 0–471–81932–8.

(5a)   Carron, L. Peter Jr (1991) *Morse code: The essential language*. the American Relay League Inc. ISBN 0–87259–035–6

(6)   Chichocki, A., Unbehauen, R. (1993). *Neural Networks for Optimization and Signal Processing*. Wiley. ISBN 0–471–93010–5.

(7)   Chen, Chi-Tsong (1999). *Linear System Theory and Design*. Oxford University Press ISBN 0–19–511777–8.

(8)   Cover, T.M., Thomas, J.A. (1991). *Elements of Information Theory*. Wiley. ISBN 0–471–06259–6.

(9)   Denbigh, Philip (1998). *System Analysis & Signal Processing*. Addison-Wesley. ISBN 0–201–17860–5.

(9a)   *DSP56000/56001 Digital Signal Processor User's Manual* (1989) Motorola Literature Distribution, Arizona, 1989. DSP56000UM/AD.

(10)   Fitch, J.P., Doyle, E.J., Gallagher Jr, N.C. (1984). Median filtering by threshold decomposition. *IEEE Trans. Acoust. Speech Signal Process.* **32**: 1183–1188.

(11)   Gallagher Jr, N.C., Wise, G.L. (1981). A theoretical analysis of the properties of median filters. *IEEE Trans. Acoust. Speech Signal Process.* **29**: 1136–1141.

(12)   Gonzales, R.C., Wintz, P. (1987). *Digital Image Processing*. Addison-Wesley. ISBN 0–201–11026–1.

(13)   Grossberg, S. (1987a). *The Adaptive Brain I*. Elsevier. ISBN 0–444–70413–2.

(14)   Grossberg, S. (1987b). *The Adaptive Brain II*. Elsevier. ISBN 0–444–70414–0.

(15)   Hillis, W.D. (1987). *The Connection Machine*. MIT Press. ISBN 0–262–08157–1.

(16)   Hinton, G.E., Anderson, J.A. (1981). *Parallel Models of Associative Memory*. Lawrence Erlbaum. ISBN 0–89859–105–8.

(17)   Holmes, P. D. (1989). *The 8-queens problem and task-scheduling – a neural net approach*. FOA, Linköping, March 1989. FOA Report D 30526-3.4.

(18)   Hopfield, J.J., Tanks, D.W. (1986). Computing with neural circuits: a model. *Science* **253**: 625–633.

(19)   Hu, M. J.C. (1963). *A trainable weather-forecasting system*. Stanford Electronics Laboratory, Stanford University, June 1963. Report No. ASD–TDR–63–636, Technical Report No. 6759–1.

(20)   Hunter, R., Robinson, H. (1980). International digital facsimile coding standards. *Proc. IEEE*, July 1980.

(21)   Lee, D.T.L., Yamamoto, A. (1994). Wavelet Analysis: Theory and Application. *Hewlett-Packard J.*, December 1994.

(22)   Lippmann, R. P. (1987). An introduction to computing with neural nets. *IEEE ASSP Mag.*, April 1987.

(23) Lynn, Paul A., Fuerst, Wolfgang (1994). *Digital Signal Processing with Computer Applications.* Wiley. ISBN 0–471–94374–6.

(24) Marven, Craig, Ewers, Gillian (1993). *A Simple Approach to Digital Signal Processing.* Alden Press. ISBN 0–904–047–00–8.

(25) McClelland, J.L., Rumelhart, D.E. (1986). *Parallel Distributed Processing, Part 2.* MIT Press. ISBN 0–262–13218–4.

(26) Miller, Gary M. (1998). *Electronic Communication.* Prentice Hall. ISBN 0–13–859828–2.

(27) Minsky, M. L., Papert, S.A. (1969). *Perceptrons,* (expanded edition 1988) MIT Press. ISBN 0–262–63111–3.

(28) Mitra, Sanjit K., Kaiser, Lames, F. (1993). *Handbook for Digital Signal Processing.* Wiley. ISBN 0–471–61995–7.

(29) Oppenheimer, Alan V., Schafer, Ronald W. (1975). *Digital Signal Processing.* Prentice-Hall. ISBN 0–13–214635–5.

(30) Orfandis, Sophocles J. (1985). *Optimum Signal Processing.* Macmillan. ISBN 0–02–949860–0.

(30a) Palm, R., Driankov, D., Hellendorrn, H. (1996). *Model Based Fuzzy Control.* Springer. ISBN 3–540-61471-0.

(31) Papoulis, Athansios (1985). *Probability, Random Variables and Stochastic Processes.* McGraw-Hill. ISBN 0–07–048468–6.

(32) Passino, K. M., Yurkovich, S. (1998). *Fuzzy Control.* Addison-Wesley. ISBN 0–201–18074–X.

(33) Pires, Jorge de Sousa (1989). *Electronics Handbook.* Studentlitteratur. ISBN 91–44–21021–3.

(34) Pohlmann, Ken C. (1989). *Principles of Digital Audio.* Howard W. Sams, Hayden Books. ISBN 0–672–22634–0.

(35) Proakis, John G. (1989). *Digital Communications.* McGraw-Hill. ISBN 0–07–100269–3.

(36) Rabiner, Lawrence R., Gold, Bernard (1975). *Theory and Application of Digital Signal Processing.* Prentice-Hall. ISBN 0–13–914101–4.

(37) Rumelhart, D.E., McClelland, J.L. (1987). *Parallel Distributed Processing, Part 1.* MIT Press. ISBN 0–262–18120–7.

(38) Schwartz, M., Shaw, L. (1975). *Signal Processing: Discrete Spectral Analysis, Detection, and Estimation.* McGraw-Hill. ISBN 0–07–055662–8.

(39) Sjöholm, S., Lindh, L. (1997). *VHDL for Designers.* Prentice-Hall. ISBN 0–13–473414–9.

(40) Specht, D. F. (1964). *Vectorcardiographic diagnosis utilizing adaptive pattern-recognition techniques.* Stanford Electronics Laboratory, Stanford University, June 1964. Technical Report No. 6763–1.

(41) Spiegel, Murray R. (1971). *Calculus of Finite Differences and Difference Equations.* McGraw-Hill. ISBN 0–07–060218–2.

(42) Stranneby, D. (1990). *Error correction of corrupted binary coded data, using Neural Networks.* KTH, Stockholm, 1990. Report No. TRITA–TTT9008.

(43) Stranneby, D. (1996). *Power and frequency assignment in HF radio networks.* Radio Communication Systems Laboratory, KTH, Stockholm, April 1996. TRITA-S3-RTS-9605.

(44) Widrow, B., Lehr, M.A. (1990). 30 years of adaptive neural networks: perceptron, madaline and backpropagation. *Proc. IEEE,* **78**(9).

(45) Widrow, Bernard, Stearns, Samuel D. (1985). *Adaptive Signal Processing.* Prentice-Hall. ISBN 0–13–004029–0.

(46) Wirth, N. (1976). *Algorithms + Data Structures = Programs.* Prentice-Hall. ISBN 0-13-022418-9.

Glossary

A brief overview of some common abbreviations and buzz-words.

A/D	Analog to digital
ADC	Analog to digital converter
ADM	Adaptive delta modulation
ADPCM	Adaptive differential pulse code modulation
AGC	Automatic gain control
A-law	*Signal companding standard used in Europe (see μ-law)*
ALU	Arithmetic logic unit
AM	Amplitude modulation
AND	*Boolean function*
ANN	Artificial neural network
ANS	Artificial neural system
APC	Adaptive predictive coding
AR	Auto-regressive
ARMA	Auto-regressive moving average
ASCII	American standard code for information interchange
ASIC	Application specific integrated circuit
ASK	Amplitude shift keying
AU	Address unit
AWGN	Additive white Gaussian noise
BCD	Binary coded decimal
BCH	Bose, Chaudhuri, Hocquenghem *(class of error-correcting codes)*
BIT	Binary digit
BSC	Binary symmetric channel
BU	Bus unit
C	*Computer programming language (see C++)*
C++	*Computer programming language (extension of C)*
CAM	Content addressable memory
CCITT	Comité Consultatif International Télégraphique et Téléphonique

CD	Compact disc
CD-ROM	Compact disc read only memory
CELP	Code excited linear predicition
CISC	Complex instruction set computer
CM	Connection machine
CMOS	Complementary metal oxide semi-conductor
CODEC	Coder–decoder
CoA	Centre of area
CoG	Centre of gravity
CoM	Centre of maximum
Compander	Compressor–expander
CRC	Cyclic redundancy check
CVSD	Continuously variable slope delta modulator
D/A	Digital to analog
DAB	Digital audio broadcasting
DAC	Digital to analog converter
DC	Direct current (*sometimes interpreted as a constant bias*)
DCT	Discrete cosine transform
DFT	Discrete Fourier transform
DM	Delta modulation
DMA	Direct memory access
DPCM	Differential pulse code modulation
DPCM-AQB	Differential pulse code modulation adaptive quantization – backwards
DPCM-AQF	Differential pulse code modulation adaptive quantization – forward
DSP	Digital signal processing
DSP	Digital signal processor
DTMF	Dual-tone multi-frequency
DVB	Digital video broadcasting
DVD	Digital video disc
ECG	Electrocardiograph
EEG	Electroencephalogram
ENIAC	Electronic numerical integrator and computer
EPROM	Erasable programmable read only memory
EU	Execution unit

EXCEL™	*Spreadsheet type calculation software by Microsoft*
FFT	Fast Fourier transform
FIFO	First in first out (*a queue*)
FIR	Finite impulse response
FM	Frequency modulation
FORTRAN	Formula translation (*old computer programming language*)
FOH	First order hold
FPGA	Field programmable gate array
FPU	Floating point unit
FSK	Frequency shift keying
GPS	Global positioning system
GSM	Groupe speciale mobile *or* global system for mobile communication
HDTV	High definition television
IEEE	The Institute of Electrical and Electronics Engineers
I²C™	(IIC) Inter IC (*simple bidirectional 2-wire bus standard developed by Philips*)
IIR	Infinite impulse response
I/O	Input/output
I/Q	In phase/quadrature phase
ISDN	Integrated services digital network
ISI	Intersymbol interference
JAVA	*Computer programming language (subset of C++)*
JPEG	Joint photographics expert group
LDM	Linear delta modulation
LED	Light emitting diode
LIFO	Last in first out (*a stack*)
LMS	Least mean square
LPC	Linear predictive coding
LSB	Least significant bit
LTI	Linear time invariant (*commonly also implies causal*)
LUT	Lookup table
LZ	Lempel–Ziv
MA	Moving average
MAC	Multiply add accumulate
MathCad™	*Calculation software by MathSoft*

Mathematica™	*Calculation software by Wolfram Research*
MATLAB™	*Calculation software by MathWorks*
MFLOPS	Million floating point operations per second
MHC	Modified Huffman code
MIMD	Multiple instruction multiple data
MIMO	Multi-input multi-output
MIPS	Million instructions per second
MISO	Multi-input single-output
MIT	Massachusetts Institute of Technology
ML	Maximum likelihood
MLPC	Multipulse excited linear predictive coding
MoM	Mean of maximum
MPEG	Moving pictures expert group
MRC	Modified Read code
MSB	Most significant bit
MSE	Mean square error
μ-law	*Signal companding standard used in USA (see A-law)*
NOP	No operation
NOT	*Boolean function*
OCR	Optical character reading
OR	*Boolean function*
OSR	Oversampling ratio
Pascal	*Computer programming language*
PAM	Pulse amplitude modulation
PC	Personal computer
PC	Program counter
PCM	Pulse code modulation
PDM	Pulse density modulation
PDS	Parallel distributed system
PM	Phase modulation
PNM	Pulse number modulation
PPM	Pulse position modulation
ppm	Parts per million
PROM	Programmable read only memory
PSK	Phase shift keying

PWM	Pulse width modulation
QAM	Quadrature amplitude modulation
QPSK	Quadrature phase shift keying
RELP	Residual excited linear prediction
RISC	Reduced instruction set computer
RLS	Recursive least square
RMS	Root mean square
ROM	Read only memory
RS	Reed–Solomon (code)
SAR	Succesive approximation register
SBC	Sub-band coding
SC	Switched capacitor
SDR	Software defined radio
S&H	Sample and Hold
S/H	Sample and Hold
SIMD	Single instruction multiple data
SNR	Signal-to-noise ratio
SP	Stack pointer
SPI	Serial peripheral interface
SSB	Single sideband
UWB	Ultra wideband
VHDL	Very high speed integrated circuit hardware description language
VLSI	Very large scale integration
VSELP	Vector sum excited linear prediction
WCDMA	Wideband code division multiple access
XOR	*Exclusive OR* (*Boolean function*)
ZOH	Zero-order hold

Index